An Introduction
to
Water Quality Modelling

An Introduction
to
Water Quality Modelling

Edited by

A. James
Department of Civil Engineering,
University of Newcastle upon Tyne, UK

A Wiley–Interscience Publication

JOHN WILEY AND SONS
Chichester · New York · Brisbane · Toronto · Singapore

Library of Congress Cataloging in Publication Data:

James, A.
  An introduction to water quality modelling.

  'A Wiley–Interscience publication.'
  Includes index.
  1. Water quality—Measurement—Mathematical models.
  I. Title.
TD367.J35   1984   628.1'0724   83–16921

ISBN 0 471 90356 6

British Library Cataloguing in Publication Data:

James, A.
  An introduction to water quality modelling.
  1. Water quality management—Simulation
     methods
I. Title
  333.91'00724   TD365

ISBN 0 471 90356 6

Typeset by Mathematical Composition Setters Ltd,
Ivy Street, Salisbury, Wilts., and printed
by Page Bros. (Norwich), Norwich.

List of Contributors

D. J. ELLIOTT   *Department of Civil Engineering, University of Newcastle upon Tyne*

R. E. FEATHERSTONE   *Department of Civil Engineering, University of Newcastle upon Tyne*

A. JAMES   *Department of Civil Engineering, University of Newcastle upon Tyne*

Contents

Preface

In the last two decades the application of modelling techniques to water quality problems has increased dramatically and many useful techniques have emerged. Unfortunately, teaching of these techniques has not received the same attention and has consequently lagged behind. The hiatus is apparent in the large number of papers and books published for the advanced practitioner compared with the dearth of simple texts for the beginner.

This book is aimed at the beginner. It is based upon a number of courses which have been held in the Civil Engineering Department of the University of Newcastle upon Tyne to introduce quality modelling to scientists and engineers who were working, or intending to work, in water pollution control.

The text is a simple guide, suitable for biologists, chemists, engineers and others without any previous knowledge of modelling or computing. The level of mathematical complexity has been deliberately restricted.

An introductory chapter explains the concepts and terminology used in simulation, optimization and computer-aided design. This is followed by two other introductory chapters dealing with computing and numerical methods. The chapter on computing includes a guide to BASIC which is the simplest programming language in common use and the language used throughout the remainder of the book.

The final introductory chapter explains the hydraulic, chemical and biological ideas which are used in formulating models of water quality.

The remainder of the book is divided into two sections dealing with the application of modelling techniques to water pollution and wastewater treatment plants.

Each chapter presents the modelling concepts and shows how these may be built into models suitable for examining frequently occurring problems. A complete listing of an example programme is included as an appendix to each chapter.

It is hoped that this book will help and encourage students and practitioners in the water quality field to understand mathematical modelling techniques.

Chapter 1

Introduction to Mathematical Modelling

A. JAMES

1.1 INTRODUCTION

The proper management of water resources even on a small scale is very difficult. There are a large number of quality criteria to be considered and in most cases the level of each criterion is the resultant of complex interactions. The situation is further exacerbated by the difficulties of any experimental approach to forecasting water quality. This has led to the growth of mathematical modelling as a means of predicting quality.

The representation of the interactions in a system by a set of equations is not a new idea. The classic work on oxygen sag by Streeter and Phelps demonstrated the possibilities. But until recently, the application of mathematical modelling was limited by the difficulty of finding analytical solutions to the equations. It is the development of computing and numerical methods of solution that has led to increasing interest in modelling.

Various kinds of mathematical models have been designed for different purposes. They may be classified under the three general headings as:

(a) Simulation.
(b) Optimization.
(c) Computer aided design.

These are discussed separately below.

1.2 SIMULATION

The creation of a mathematical model which simulates water quality changes can be carried out on different levels. The aim in simulation should always be for the maximum simplicity consistent with the required degree of accuracy and detail.

The steps in creating a simulation model are summarized in Fig. 1.1. This shows the definition of the problem as being the first stage. The objectives must be clearly stated if a simple but realistic model is to be obtained, if possible in numerical terms.

The second stage in model building is to review the current theory relating the various parameters of the system. It is worthwhile looking at alternative relationships or forms of relationships at this stage. In particular there is the choice between causal and statistical relationships.

Depending upon the current state of the theory the formulation of equations may or may not present some difficulty. Where there are no established relationships some laboratory or pilot-plant work may be needed together with some statistical analysis to find the form and reliability of the connection. Care should be taken to avoid over-complication of the model by eliminating all relationships which do not significantly affect the result.

In choosing the relationships to include, it is best to adopt an hierarchical approach. Including first those relationships which are essential and then using other equations to modify the relationship for secondary factors. For example, in attempting to represent algal growth, it is possible to think of a number of factors such as light, temperature, nutrients, pH and predators, all of which influence the growth rate. Having first considered whether it is essential to include all of these, they can then be sorted out into primary and secondary relationship, viz.:

Primary relationships

Algal growth is a function of light intensity.
Algal growth is related to nutrient availability.
Predation if important is a primary factor.

Secondary relationships

Temperature correction needed for growth and respiration.
pH correction needed for growth.
The arrows indicate the general sequence of steps in model building but some iteration is usually necessary since decisions in the later steps may cause changes in earlier steps.

Creating the structure of the model is the most difficult stage of model building particularly with more complex models. It is generally advisable to begin by identifying the large sub-divisions of a model and fitting these together in a diagrammatic manner. This helps to establish the overall flow of information through the model and highlights the difficulties of interfacing the sub-divisions. It may be necessary in some models to have more than one attempt at choosing the sub-divisions. An example of the sub-division process is given in Fig. 1.2.

Once the overall structure has been established the second stage of the model structuring is to fill in the details of input, output, calculation and decision-making; this can be done in various ways but it usually involves preparing flow

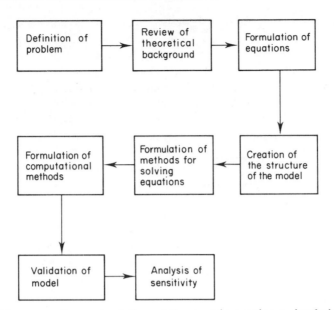

Fig. 1.1   Diagrammatic representation of the steps in creating a simulation model

charts either as a block diagram or as an electrical analogue. These alternatives may be summarized as follows:

(a) A generalized block diagram given the succession of steps in the computation with iteration loops, decision steps, etc. There is no standard format for this type of diagram, but the different shapes of boxes may be used to indicate input, calculation, decision and output steps.

(b) Alternatively, a model may be represented in the form of an electrical analogue circuit diagram as shown in Fig. 1.4 where each operation of input and calculation is represented by a conventional symbol. The symbols and their meaning are shown in Fig. 1.5. This is a much more detailed approach and is not suitable for representing the overall structure of complex models. But provided that the model can be broken down into a number of sub-sections, then the analogue diagrams for each can be drawn.

This diagrammatic representation helps in identifying sub-sections of the system and helps in classifying them. For example, in Fig. 1.2 the various sources of and sinks for oxygen are grouped together and are differentiated from other parts of the system such as the physical and hydraulic characteristics. It also helps in establishing the paths for information flow and helps to avoid incorrect sequences in the chain of calculations.

This approach has the advantage of giving an exact step by step guidance when the program is being written.

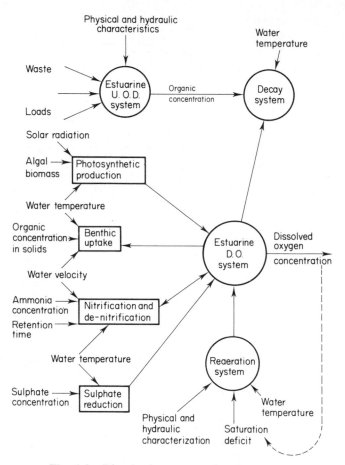

Fig. 1.2   Dissolved oxygen regime in an estuary

The numerical methods for solving equations are discussed in Chapter 3. There is no obligation to use numerical methods if a suitable analytical solution is available but in most models the application of analytical methods is very limited.

The computation may be carried out on any suitable machine from a calculator upwards. For more complex models and where many alternatives need to be explored, the use of a computer is essential. The form of data inputs and outputs and the choice of language are largely dictated by the facilities available. If the modelling is particularly complex a considerable saving of programming effort can result from the use of a high-level language like CSMP (Continuous System Modelling Program). In the engineering field, most models are programmed in FORTRAN whereas in pure science ALGOL

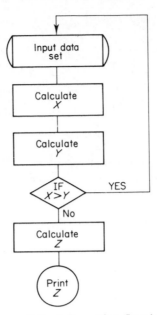

Fig. 1.3 Block diagram of the information flow in a simulation model

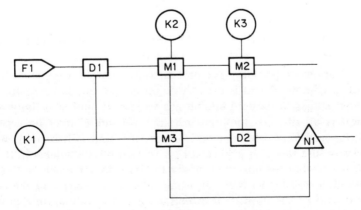

Fig. 1.4 Analogue diagram of information flow

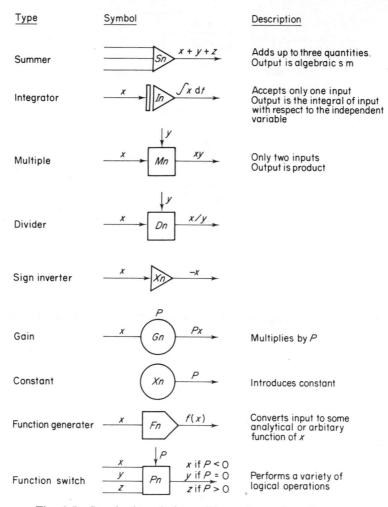

Type	Symbol	Description
Summer	Sn $x + y + z$	Adds up to three quantities. Output is algebraic s m
Integrator	x In $\int x \, dt$	Accepts only one input Output is the integral of input with respect to the independent variable
Multiple	y x Mn xy	Only two inputs Output is product
Divider	y x Dn x/y	
Sign inverter	x Xn $-x$	
Gain	P x Gn Px	Multiplies by P
Constant	Xn P	Introduces constant
Function generater	x Fn $f(x)$	Converts input to some analytical or arbitary function of x
Function switch	P x y z Pn x if $P < 0$ y if $P = 0$ z if $P > 0$	Performs a variety of logical operations

Fig. 1.5　Standard symbols used in analogue flow diagram

and Pascal are more popular. The best introductory language is BASIC. It is described in Chapter 2 and is used throughout the remainder of the book.

The final stages in model building are validation and sensitivity analysis. Validation is an obvious requirement since no model may be accepted as representing a system without suitable proof. It is normally carried out in two steps. Firstly, the model is calibrated, i.e. values of coefficients in the model are adjusted so that the output from the model is in agreement with observation. Secondly, the model is re-run using other input data and the output is compared with observation. It is essential for a real validation that the data sets for calibration and testing are kept separate.

Sensitivity analysis is used to determine the accuracy required in measuring or estimating the various coefficients in the model. It is carried out by running the model several times, each time incrementing the value of the particular coefficient and examining the effect that this has upon the output.

In this way the relative importance of the coefficients is made apparent and a corresponding allocation of effort can be made in estimating their numerical value.

1.2.1 Simulation example

An example of a simulation problem is illustrated in Fig. 1.6. It shows a proposal to discharge the effluent from a coke-oven works into a stream some distance upstream of the abstraction point for a municipal water treatment plant. The problem is to decide on the consent conditions to be imposed on the discharge.

(a) *Formulation of objectives.* These need to be as precisely stated as possible, preferably in quantitative terms.

Coke-oven effluents contain various pollutants notably ammonia, cyanide, phenol and thiocyanate. Of these the most critical for the water treatment plant is phenol and the significant level is $0.001\,\text{mg l}^{-1}$.

(b) *Review of theory.* The concentration of phenol in the river at the abstraction point will be influenced by various factors, mainly
— initial dilution by freshwater flow;
— dispersion due to molecular and eddy diffusion;
— breakdown of phenol by bacterial activity;
— dispersion and breakdown rates are both dependent on temperature and freshwater flow.

At this stage, it is useful to consider whether it is essential to represent all of processes, e.g. it may be decided that dispersion and/or breakdown are insignificant and may therefore be neglected. In this example, it has been decided to omit longitudinal dispersion. The dispersion effect will therefore be included in with the decay and it must be appreciated that the decay coefficient is a lumped parameter.

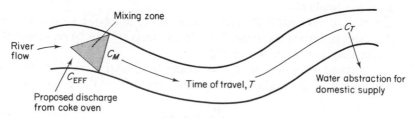

Fig. 1.6 Example of a water quality simulation problem

(c) *Formulation of equations.* Following on from (b) the chosen processes need to be formulated as equations. The initial dilution due to mixing with the freshwater flow can be stated as a mass balance in words

$$\text{Concentration of phenol after initial dilution} = \frac{\text{Upstream flux of phenol} + \text{Effluent flux of phenol}}{\text{upstream flow} + \text{effluent flow}}$$

or in symbols

$$C3 = \frac{(Q1 * C1 + Q2 * C2)}{(Q1 + Q2)} \tag{1.1}$$

and the decay process by

$$\text{Concentration of phenol after decay} = \text{Concentration after dilution} * \text{Negative exponential of decay rate * time}$$

$$C4 = C3 * \exp(-K * T) \tag{1.2}$$

and the time of travel, T

$$\text{Time of travel} = \frac{\text{Volume of reach}}{\text{Upstream flow} + \text{Effluent flow}}$$

$$T = \frac{V}{(Q1 + Q2)}$$

where

$Q1$ = upstream flow
$C1$ = upstream concentration pf phenol
$Q2$ = effluent flow
$C2$ = phenol concentration in effluent
K = decay coefficient

The above relationships are only meant to be examples. It may be necessary to improve on some of these, particularly the calculation of the time of travel. A sensitivity analysis would be used to determine the accuracy required.

(d) *Creation of the model structure.* In a model involving so few relationships this is a trivial exercise. The model structure can be represented by a flow diagram as shown in Fig. 1.7.

(e) The method of solution in this example is analytical since with the elimination of dispersion there is only one differential equation:

$$\frac{dC}{dt} = -KC \tag{1.3}$$

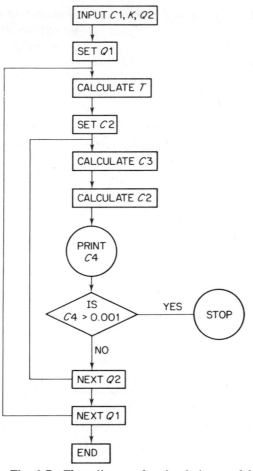

Fig. 1.7 Flow diagram for simulation model

which is easily integrated to give

$$Ct = Co \exp(-Kt) \qquad (1.4)$$

(f) Since very few steps are involved a simple program can be produced for the model

Line No.	Instruction	Explanation
10	INPUT K, V	Introduces fixed values for decay coefficent and volume of reach
20	$Q1 = 5.0$	Sets minimum freshwater flow

30	For Q2 = 2 to 5 STEP 1	Iterative loops for flow of effluent at levels of 2, 3, 4 and 5
40	For C1 = 1 to 101 STEP 10	Iterative loop for phenol concentration in effluent at levels of 1, 11, 21 etc. up to 101
50	C2 = (C1 × Q2)/(Q1 + Q2)	Calculate phenol concentration after mixing
60	T = V/(Q1 + Q2)	Calculate retention time in the reach
70	C3 = C2* EXP(−K*T)	Calculate concentration at end of reach
80	PRINT C3	Outputs phenol concentration
90	IF C3 0.001 THEN 120	Jumps to end if critical concentration is exceeded
100	NEXT C1	End of concentration loop
110	NEXT Q2	End of flow loop
120	END	End statement

(g) The validation of the model will depend upon local possibilities. Ideally one would like to test the output from the model against the actual discharge for several effluent flows or concentrations. This can most easily be simulated by a gulp discharge of phenol at the site of the proposed discharge or more elaborately by arranging a temporary pipeline. If field experiments are not possible with phenol it may be possible to check experimentally on the rate of decay and the time of travel.

(h) The sensitivity analysis should be used to determine the importance of the decay coefficient and time of travel relationships, since these are the two inputs which are most difficult to establish. Some analysis may also be required on two other items, viz.:

(i) Longitudinal dispersion,
(ii) Incomplete mixing.

1.2.2 Simulation terminology

A number of special terms are used to characterize simulation models. The ones in common usage are described below:

(i) Steady-state—models where the system is not varying with time. The inputs to the model are fixed and consequently the system eventually reaches some equilibrium condition. Confusion may arise with models of batch systems because the inputs are fixed but no steady-state is achieved. Such models are pseudo-dynamic models—see Chapter 5 for discussion of Streeter–Phelps batch model of dissolved oxygen in a river. Steady-state models have various attractions—notably simplicity and small data requirements.

(ii) Dynamic—models where the system is varying with time due to changing conditions or due to changing inputs. Dynamic models are more difficult to formulate and require larger amounts of data but in many circumstances like estuaries can be justified. They also require much more computing time which gives rise to problems with smaller computers.

(iii) Deterministic—models where there is a fixed relationship between the input and output. All the relationships in such a model are rigidly determined and the input and output are similarly assumed not to be subject to error.

(iv) Stochastic—models which allow for random fluctuation in the system due to variations in the input parameters, variations in the state of the system coefficients or errors in the output measurements. The variations are introduced by fixing the mean value of a parameter or coefficient together with the standard deviation. Data requirements for such a model are obviously greater than the equivalent deterministic model.

1.3 OPTIMIZATION

Optimization is a group of mathematical techniques which are used to obtain the best solution from a range of possibilities. The best solution may be a minimum, e.g. a least cost solution, or a maximum, e.g. largest improvement for a given investment. Whatever the technique employed the problem needs to be expressed in the same way, as follows:

1. *Objective function:*
   The expression to be optimized must be written in the form of an equation e.g.

$$C = \sum_{i=1}^{n} W_i B_i \qquad (1.5)$$

Total cost = the sum for all plants $\left(\begin{array}{cc} \text{The cost per} & \text{The number of} \\ \text{unit improvement} \times & \text{units of improvement} \\ \text{at a plant } i & \text{requires at plant } i \end{array} \right)$

2. *Constraint equations:*
   The various boundaries which are imposed upon the solution must also

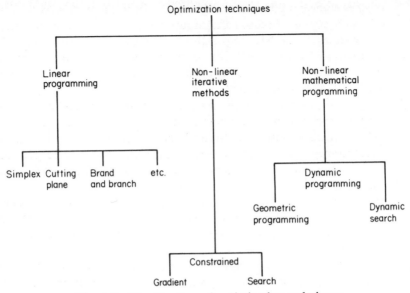

Fig. 1.8 Classification of optimization techniques

be expressed as equations in the form

$$A_1X_1 \pm A_2X_2 \ldots A_nX_n \leqslant Y1$$

$$B_1X_1 \pm B_2X_2 \ldots B_nX_n \leqslant Y2$$

These equations may then be manipulated by a variety of methods. The full description of these methods is outside the scope of this chapter but the methods may be briefly categorized as shown in Fig. 1.8.

An example of linear programming is described below and examples of a search technique and dynamic programming are described in Chapter 14.

1.3.1 Optimization example

An effluent re-use scheme is being examined and it is designed to maximize the profit. Two qualities of effluent are required X_1 and X_2. These can be obtained subject to the constraints that:

$$2X_1 + 4X_2 \not> 20$$

$$2X_1 + 2X_2 \not> 12$$

$$4X_1 \qquad \not> 16$$

The constraint equations contain inequalities. This may be overcome by introducing u_1, u_2 and u_3 as slack variables and re-writing the equations as

$$u_1 + (2x_1 + 4x_2) = 20$$
$$u_2 + (2x_1 + 2x_2) = 12$$
$$u_3 + (4x_1 + OX_2) = 16$$

The final row of the matrix is the objective expressing the current value of the profit. Thus if the basic equations are:

Objective function:

Maximize

$$C = 2x_1 + 3x_2$$

Subject to the constraints that

$$2x_1 + 4x_2 \leqslant 20$$
$$2x_1 + 2x_2 \leqslant 12$$
$$4x_1 \leqslant 16$$

these may be written as

	x_1	x_2	
u_1	2	4	20
u_2	2	2	12
u_3	4	0	16
$-C$	2	3	0

The initial solution of the simplex as represented by the above table is that

$$x_1 = x_2 = 0$$
$$u_1 = 20$$
$$u_2 = 12$$
$$u_3 = 16$$
$$C = 0$$

This is the basic feasible solution from which we are seeking to improve. The difficulty in carrying out the improvement is in manipulating.

From the objective function it is apparent that more profit is obtained from x_2 rather than x_1. By rewriting the first constraint equation we can obtain

$$x_2 = 5 - 0.5x_1 - 0.25u_1$$

and by substitution in the other equations we get:

$$u_2 + x_1 - 0.5u_1 = 2$$

$$u_3 + 4x_1 = 16$$

$$-C + 0.5x_1 - 0.75u_1 = -15$$

so the simplex matrix becomes

	x	u_1	
x_2	0.5	0.25	5
u_2	1	-0.5	2
u_3	4	0	16
$-C$	0.5	-0.75	-15

The same transformation from matrix 1 to matrix 2 can more simply be effected by the following rules:

1. Identify the pivot element. This is the element which stands in the row and column of the two variables which are to be interchanged.
2. The pivot element must be selected from the column whose last row contains the largest positive element for maximization or the lowest element for minimization.
3. Within the column the pivot element is the one with the lowest production ratio i.e.

$$\frac{\text{column element}}{\substack{\text{corresponding element} \\ \text{in last column}}}$$

4. The pivot element is replaced by the reciprocal, R that relates the amount of time required at each treatment plant for unit production of the two effluent qualities. The constraints to the system can be entered as a final column—maximum amount of time available at each treatment plant.
5. The other elements in the row of the pivot are calculated by multiplying element by R.
6. The other elements in the column of the pivot are calculated by multiplying the previous element by R.
7. To complete the rest of the matrix proceed by column. In each incomplete column there is already one element E. The remaining elements of the column are calculated by

$$\text{New element} = \text{Old element} - \left(\begin{array}{l} \text{Ex element in old table in the row of} \\ \text{old element and column of the pivot} \end{array} \right)$$

Using these rules the pivot in the existing matrix may be identified as x_1/u_2 and the new table becomes:

	u_2	u_1	
x_2	-0.5	0.5	4
x_1	1	2	8
u_3	-4	2	8
$-C$	-0.5	-0.5	-16

Since all the bottom row elements are negatives an optimal solution has been reached where $x_1 = 2$ and $x_2 = 4$ with a total profit of 16.

1.3.2 Application of Optimization

Optimization has been used extensively in water resource management mainly for resource allocation. An example is discussed in Chapter 5. Application of optimization in other fields of water pollution control such as waste treatment plant designs have been limited by the lack of suitable simulation models to represent processes like biological filtration, sedimentation, etc. An example of optimization in treatment plant design is discussed in Chapter 14 but clearly further improvements in the simulation of unit processes are required if the developments in optimization are to be of benefit.

1.4 COMPUTER-AIDED DESIGN

This branch of mathematical modelling has received little attention in public health engineering and interest in CAD has been concentrated in structural design. However, the potential benefits of CAD become obvious when one considers the proportion of time devoted to the various stages of design e.g.

Select process type	5%
Process design	10%
Structural design	15%
Calculation of quantities	25%
Preparation of drawings	45%

Simulation and optimization can help with the choice of process and the process design but it is in the later stages particularly the preparation of drawings that the greatest savings in labour can be made.

The steps in computer-aided design may be briefly outlined as follows:

1. Simulation model of unit process to determine dimensions of the unit(s).
2. Input of coordinates of site. Determine location of unit(s).

3. Structural design of unit.
4. Calculation of coordinates for unit.
5. Calculation of quantities of concrete and steel.
6. Calculation of quantities of excavation, shuttering, etc.
7. Preparation of drawings.

Computer graphics are discussed in Chapter 2 and an example of computer-aided design is given in Chapter 10.

1.5. DATA COLLECTION

In attempting a modelling exercise to help solve a water quality problem there is inevitably a requirement for data. This requirement should be examined during the course of the model formulation. There are two points to bear in mind:

(a) The type of survey required for model building is fundamentally different from that for routine surveillance. The latter is essentially long-term monitoring in a sparse manner whereas model building generally requires an intensive data gathering over a limited period.
(b) Data (or the lack of them) are likely to determine the success of the modelling exercise. The time and effort needed in the collection of data for calibration and validation is likely to far exceed the time spent in developing the model. In fields like estuarine modelling the ratio of times may be greater than 10 : 1 (data collecting : modelling)

1.6 BIBLIOGRAPHY

Some useful books on water quality modelling are listed below.

Thomann, R. V. (1972). *Systems Analysis and Water Quality Management*. McGraw-Hill, New York.
Rich, L. G. (1973). *Environmental Systems Engineering*. McGraw-Hill, New York.
James, A. (ed.) (1978). *The Use of Water Quality Models in Water Pollution Control*. John Wiley & Sons, Chichester.
Franks, R. G. E. (1972). *Modelling and Simulation in Chemical Engineering*. John Wiley & Sons, New York.
Vemuri, V. (1978). *Modelling of Complex Systems*. Academic Press, London.
Biswas, A. K. (ed.) (1976). *Systems Approach to Water Management*. McGraw-Hill, New York.
Gordon, G. (1969). *System Simulation*. Prentice Hall, New York.
Morley, D. A. (1979). *Mathematical Modelling in Water and Wastewater Treatment*. Academic Press, London.

Chapter 2

Introduction to Computing

A. JAMES

2.1 INTRODUCTION

All but the simplest of mathematical models require the use of a computer but it is important that the computational aspects of modelling are kept in perspective as computing can all too easily become a way of life.

The aim of computing is to obtain as easily and efficiently as possible the output in a suitable form. There needs to be a balance between the time and effort in programming and the efficient use of the computer. Sophistication in programming is not necessarily a virtue.

The first choice to be made in computing is the type of computer (hardware). This is determined by consideration of model size (amount of data storage) cost, etc. The rapid developments in computer technology make advice rapidly out of date but the general trend in hardware is towards smaller machines and much modelling work can be carried out on 16K or 32K machines. Large estuarial and marine models obviously require greater storage and processing power. Careful consideration of the computing requirements needs to be made based not only on modelling but also on other computing needs.

Another consideration is the length of time required for the computer to carry out the calculations. Solution of large network problems may take several hours on a microcomputer and only seconds or minutes on a large machine.

The other aspect of choice is in the software, i.e. the purchase of commercial programs. A large and expanding industry has developed which produces software packages (i.e. groups of programs) to carry out standard theoretical operations e.g. regression, matrix manipulation, etc. or specific applied techniques, e.g. pipe network analysis. Such packages can be extremely useful and their availability may influence the choice of computer.

2.2 PROGRAMMING IN BASIC

A description of a high-level computer language such as FORTRAN is beyond the scope of the book. The following notes provide an introduction to BASIC.

This has become one of the most widely used languages because of its simplicity and also because it allows an easy progression to FORTRAN.

The various stages in formulating a mathematical model have already been described in Chapter 1. Writing the program is carried out after the production of the flow diagram since this acts as a step by step guide.

The different aspects of programming may be briefly described as follows:

(a) Signing on/off—This operation is not necessary on a single-terminal machine, but where several users and terminals are being run on a larger machine it is necessary to establish to the machine the identity of the user. The format of doing this is rather variable but generally consists of three parts, e.g.

$$\begin{array}{cccc} \text{HELLO} & - & \text{C600,} & \text{BLANK} & \left(\begin{array}{c} \text{Press} \\ \text{return key} \end{array} \right) \\ \uparrow & & \uparrow & \uparrow & \\ \text{COMMAND} & & \text{IDENTITY} & \text{PASSWORD} & \\ & & \text{OF USER} & & \end{array}$$

Computers are often fussy about the details such as spaces or commas or hyphens between the parts, but provided the current format is observed an acknowledgement will be received. The note at the end about the return key raises the fundamental point that communication with the computer is a two-way operation. Having switched on, the machine will wait for instructions (commands) from you or may prompt you to give instructions. Once you have given an instruction, pressing the return key transfers command to the machine so that it may execute the instruction. Once the instruction such as signing on is executed, then command is returned to you. Often the computer will indicate the return of command to the operator by message such as READY or DONE.

Signing off is generally easier than signing on. The point to remember with both single and multiple user machines is that if any programs that have been created are not stored before the sign off, then they will be lost. There are two kinds of memory within a computer—temporary which can be used while the user is signed on (or while the machine is switched on) or permanent which requires some separate instructions—see storage.

(b) Creating a file—Each program is written as a separate file. There is some variation in the method of naming programs but all files should be given a name. A file may contain a program, or data or both.

To create a file, the instructions or data are entered as a series of lines each one being given a number. When the computer executes a program it does so in numerical sequence. The line numbers are for reference and to determine the sequence of operation. It is therefore best to leave intervals of 10 in the numbering to allow the insertion of intermediate lines.

At the end of file it is often necessary to have an END statement.

Having completed work on one program, it is important to indicate to the computer by commands such as SCRATCH or NEW that work is now beginning on another file. Otherwise the information will become garbled.

Once a file has been created and stored it may be retrieved at any time. Where programs are being stored on tape or disc the commands LOAD and DLOAD will be needed (see storage).

(c) Input—A numerical value may be assigned to any symbol by an instruction e.g. LET $X = 10$. Alternatively, the command INPUT X will cause a value to be asked for each time the program is run.

A third alternative is to use the command READ. This will cause the computer to search for a corresponding statement entitled DATA and will take a value from there. Where the READ occurs within a loop the computer will take successive values from the DATA, e.g. the example used in Chapter 1 could be rewritten as

$$30 \quad \text{FOR } N = 1 \text{ to } 4$$
$$40 \quad \text{READ } Q2$$
$$\cdot$$
$$\cdot$$
$$\cdot$$
$$105 \quad \text{DATA } 2,3,4,5$$
$$110 \quad \text{NEXT } N$$

Care is obviously required where several READs are employed in a program to ensure that the data is entered in the correct sequence. The command RESTORE can be used to reset the data to the beginning.

(d) Calculation—The mathematical operations are set out as instructions to the computer using the symbols $+$, $-$, $*$, $/$ plus a variety of additional facilities such as $\uparrow$ for raise to the power, LOG for logarithm, etc.

Calculation involving indices are carried out first followed by multiplication and division and thirdly addition and subtraction. Operations of equal priority are performed in sequence beginning at the left. Brackets may be used to give priority to certain steps. The equation

$$C3 = \frac{C1 * Q1 + C2 * Q2}{Q1 + Q2}$$

should be written as

$$\text{LINE } 40 \quad C3 = (C1 * Q1 + C2 * Q2)/(Q1 + Q2)$$

Brackets may also be used to prevent two arithmetic functions coming together.

Repetitive calculations are carried out using a combination of FOR and NEXT. The example shown in (c) above would cause all lines between 30 and 110 to be executed four times. In this example, the variable N has no physical significance, whereas the program written in Chapter 1 used real variables. Where the latter are used it is often helpful to be able to specify the step size (see the program in Chapter 1).

(e) Logic—Decisions in programs may be taken using IF and THEN e.g. line 90 in the program in Chapter 1

$$90 \quad \text{IF } C3 > 0.001 \quad \text{THEN } 120$$

will cause the program to leave the iteration loop and stop if the phenol concentration exceeds $0.001 \, \text{mg} \, 1^{-1}$.

When a program has two branches due to an IF—THEN statement, it is often useful to also use GO TO. For example, in CAD of sedimentation tanks there is a choice of circular or rectangular tanks. The program therefore divides at line 200.

200   IF $T = 1$ THEN 300 (indicates circular tank: $T = 0$
         is rectangular).

200 ⎫
    ⎬ Calculation for rectangular tank.
280 ⎭

290   GO TO 400 (Returns to main program after calculation
         for circular tank.)

300 ⎫
    ⎬ Calculations for circular tank.
390 ⎭

400   Main program.

Where a particular set of calculations is required at several steps in a program, the command employed is GOSUB which directs the program to the subroutine required. At the end of the sub-routine, the command RETURN causes a return to the main program at the next line after GOSUB.

(f) Output—The value of any of the variables, final or intermediate, may be obtained by using PRINT (or sometimes WRITE). Where more than one variable is being printed out, it is useful to be able to identify the variable. This can be done as follows:

$$100 \quad \text{PRINT } ``C2 = ", \quad C2$$

will print out $C2 = $ *and the current numerical value of* $C2$. Tabular and graphical output is possible with many installations. A study of the local manuals and a little practice will usually show how this may be obtained.

(g) Running—Once a program has been completed it may be executed using the command RUN. Often the first attempt to run a program will reveal errors in the programming. Some of these will cause execution to stop and an error message to be given e.g.

FOR without NEXT in line 30

In these cases it is a simple matter to correct the program by inserting a new line.

More subtle errors in the calculation steps may lead to spurious output and it is wise to test the program against known results for one run. Where spurious output is affecting the execution of the program, it may be stopped by pressing the appropriate key (often BREAK) so that the editing may be carried out.

(h) Storage—whilst the terminal or the computer is switched on, most of the storage (i.e. memory) is available to the programmer. But any permanent storage of programs or data must be arranged on tape or disc before signing off. The usual command is SAVE followed by the name of the file. The program will then be stored permanently and can be recovered at any time by LOAD (on personal computers) or GET on large main frame machines.

Computers vary in the details of the BASIC that they accept and the above notes should be used in conjunction with the appropriate instruction manual. The appendix at the end of this chapter describes most common commands in BASIC.

2.3 COMPUTER GRAPHICS

The idea of using computers to produce drawings has been in existence since the early 1960s but this aspect of modelling has been very slow to develop in the design of wastewater treatment plants. However, as previously pointed out, the potential savings in the preparation of drawings by computers can be considerable.

Whilst simulating and optimization are useful in the process design, they do not make any useful contribution to structural design or calculation of quantities or preparation of drawings. Since these are the stages which involve most expenditure of time, it is useful to be able to carry out the complete design process by computer.

The structural design and calculation of quantities are relatively straightforward operations. Appropriate algorithms are described in Chapter 10.

The preparations of drawings is a more complex operation via the computer but the techniques are now well established and a large number of software packages are available. In essence, only a limited number of operations are involved. These may be briefly outlined as follows:

(a) Tabulation of coordinates—The location of each point in a design can be specified in terms of a system of two dimensional or three dimensions coordinates. These coordinates are calculated on the basis of the process design (size and shape of the unit), and the structural design (thickness of walls, etc.). It may, in more sophisticated applications, take account of the site layout and topography so that it is possible to locate units and arrange their interconnection.

The coordinates (relative or absolute) are then tabulated for each significant point.

(b) Preparations of drawings—There are only two basic routines involved—Move and Draw.

These are best illustrated by reference to Fig. 2.1. The instructions to the plotter would be as follows:

1	MOVE	to	1
2	DRAW	to	2
3	DRAW	to	3
4	DRAW	to	4
5	DRAW	to	5
6	DRAW	to	9
7	DRAW	to	10
8	DRAW	to	6
9	MOVE	to	7
10	DRAW	to	8

The coordinates of all ten points being obtained from the table already compiled by the computer.

(c) Modification of drawing—In some situations, a variety of designs may be considered and it is most convenient to do this using a visual display unit. The various possibilities may be examined on the screen and only one or two alternatives actually executed on the plotter.

This interactive technique is particularly useful where detailed connection of units and layouts are being considered.

More advanced programming techniques are available for scaling, translation, rotation and clipping. For example, point P_1 with coor-

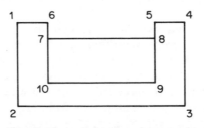

Fig. 2.1 Preparation of drawings by computer graphics

dinates X_1 Y_1 is being transformed either by translation or scaling to the point P_2 with coordinates X_2 Y_2. Then the scaling and translation operations can be expressed as 1 and 2 respectively.

$$(X_1 \ Y_1 \ 1) \begin{bmatrix} Ts_x & 0 & 0 \\ 0 & Ts_y & 0 \\ 0 & 0 & 1 \end{bmatrix} = (X_2 \ Y_2 \ 1) \qquad (2.1)$$

$$(X_1 \ Y_1 \ 1) \begin{bmatrix} 1 & 0 & 0 \\ 0 & 1 & 0 \\ Tt_x & Tt_y & 1 \end{bmatrix} = (X_2 \ Y_2 \ 1) \qquad (2.2)$$

where

Ts_x and Ts_y are scaling transform in the x and y direction Tt_x and Tt_y are translation transforms in the x and y direction.

Computer graphics are not limited to large main-frame installations and surprisingly good graphics packages are available for desk-top machines. But it must be appreciated that the preparation of working drawings requires a high precision plotter which can only be associated with larger computers.

2.4 BIBLIOGRAPHY

Books on computing tend to date rapidly due to continual changes in the machines and in programming languages. Most computers have a guide to the commands and facilities available for that specific machine. These may be used as background reading for programming in BASIC or in FORTRAN.

The following books may provide some useful ideas.

Corlett, P. N. (1968). *Practical Programming*. Cambridge University Press, Cambridge.
Kemeny, J. G. and Kurtz, T. E. (1967). *Basic Programming*, John Wiley & Sons, New York.
McCracken, D. D. (1967). *FORTRAN with Engineering Applications*, John Wiley & Sons, New York.
Monro, D. M. (1974), *Interactive Computing with* BASIC, Edward Arnold, London.
Sherman, P. M. (1970). *Techniques in Computer Programming*, Prentice Hall, New York.

APPENDIX

List of BASIC Commands

APPEND	Used to add additional data to the end of a file
BACKUP	Used to duplicate a file on another disc

CLOSE	Closes a file on the printer or other peripheral
*CLR (Clear)	Sets all variables and strings to 220 and frees all array space
CMD (Command)	Used together with OPEN to direct material to the printer—See OPEN
COLLECT	Gets rid of improperly OPENED files from disc
CONCAT (Concatenate)	Joins files together
CONT (Continue)	Allows a program to proceed after STOP key has been pressed or after an END
COPY	Copies a file within a disc unit
DATA	List of numeric data and string contents that are used by READ statements
DCLOSE	Closes disc files
DEF FN	Defines a function
DIM	Specifies storage space for arrays
DIRECTORY	Gives a listing of programs stored on a disc
DLOAD	Loads a program from disc
DOPEN	Opens a disc file
DSAVE	Stores a file on disc
END	Terminates a program and returns the computer to the command mode
FOR—NEXT	Used for beginning an iteration loop and end it
GET	Reads a character from a file into a variable (It is also used in some installations for retrieving programs from store.)
GOSUB—RETURN	Directs the computer to a sub-routine and returns it back to the main program
GO TO	Directs the computer to a particular place in the program
HEADER	Used to name or re-name a disc
IF—THEN or IF—GO TO	Used as decision step in a program
INPUT	Allows data to be entered from the keyboard as the program is executed
LET	An optional command for assigning a value to a variable. In most forms of BASIC the LET may be omitted
LIST	Causes a display of the current program
LOAD	Causes a file to be loaded from tape
NEW	Used to clear the memory of a file prior to beginning work with a new file

* Some commands are used in the abbreviated form. The full version is brackets.

ON—GOSUB or ON—GO TO	Used as a multiple decision step. The expressed expression may take several values and the computer is accordingly directed to the appropriate part of the program
OPEN	Establishes an input/output channel with another device such as a printer
PEEK	Displays on the screen whatever is stored in a particular location
POKE	Stores a byte in a particular memory location
PRINT	Causes data, etc. to be outputted
READ	Used in combination with a DATA statement to read in the value of variables
REM (Remark)	Used to insert comments into a program in the form of non-executable statements
RENAME	Changes the name of a disc file
RESTORE	Allows DATA statements to be re-read from the beginning
RUN	Causes the execution of the current program
SAVE	Stores the current program on tape
SCRATCH	Removes a disc file. Used in some versions of BASIC in place of NEW
STOP	Used to terminate the execution of a program and return to the command mode
VERIFY	Compares the contents of a program stored on tape or disc with the version in the current memory

Chapter 3

Introduction to Numerical Methods

D. J. ELLIOTT

3.1 INTRODUCTION

The process of formulation of a water quality model may be broken down into
a discrete number of steps, the first two of which are:

(1) Identify the physical, chemical and biochemical laws which govern the
    system under consideration.
(2) Express these laws in a precise mathematical form.

Frequently, in water-quality models, the equations produced in step (2) are
not amenable to direct analytical solution and simplifying assumptions are
made to reduce the complexity of the problem. Often, the equations are
reduced to a level at which an analytical solution is possible. However, even
with a simplified formulation it may be more convenient to use a numerical
rather than an analytical approach to finding a solution.

This chapter introduces some of the numerical techniques which may, with
the aid of a computer, form the basis of a water quality model.

3.2 SYSTEMS OF SIMULTANEOUS LINEAR EQUATIONS

Linear equations are used to model many different types of system. The set of
equations may be the direct result of the way in which the model is formulated
such as in the large economic models with many variables or the equations may
be developed indirectly through the use of numerical analysis to solve problems. For example, the solution of differential equations by finite difference
methods and statistical regression analysis both produce systems of linear
equations which require solution.

3.2.1 Linearity and the existence of solutions

A linear equation contains in each term only one variable and each variable appears only to the first power

$$x + y - 3z = 1 \quad \text{is linear} \tag{3.1}$$

$$xy + 3z - 9 = 9 \quad \text{is not linear} \tag{3.2}$$

$$x^2 - y - z = 11 \quad \text{is not linear} \tag{3.3}$$

A solution exists to a system of linear equations if on substituting a set of values for all variables in the system, all equations are satisfied simultaneously.

$$x + y + z = 9 \tag{3.4}$$

$$2x - 3y + z = 1 \tag{3.5}$$

$$x - y + 3z = 7 \tag{3.6}$$

This set of equations has the solution $x = 4$, $y = 3$, $z = 2$.

It is usually not possible to say without detailed examination of the equations if there is a solution and if it is unique.

Three possibilities exist. Consider Fig. 3.1

Line A is represented by $3y - 2x + 1 = 0$
Line B is represented by $y + 2x - 5 = 0$
Line C is represented by $y + 2x - 12 = 0$

(1) The system

$$\left.\begin{aligned} 3y - 2x + 1 &= 0 \\ y + 2x - 5 &= 0 \end{aligned}\right\} \tag{3.6}$$

has the solution $x = 2$, $y = 1$ which is unique.

(2) The system

$$\left.\begin{aligned} y + 2x - 5 &= 0 \\ y + 2x - 12 &= 0 \end{aligned}\right\} \tag{3.7}$$

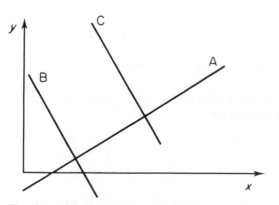

Fig. 3.1 Graph of three simultaneous equations

has no solution. The lines are parallel and because they never meet they have no solution.

(3) The system

$$y + 2x - 5 = 0 \atop 2y + 4x - 10 = 0 \Bigg\} \tag{3.9}$$

has an infinite number of solutions because the equations describe the same line and any point on the line is a solution.

Systems (2) and (3) are said to be singular. It is necessary to test for singularity which may be done directly using the determinant of the system or indirectly through the particular method of solution adopted.

Singularity of a system of equations is indicated if the determinant of the system coefficients is zero.

3.2.2 Determinants

The solution to the following system of equations

$$a_1x + b_1y = c_1 \quad \text{(a)} \atop a_2x + b_2y = c_2 \quad \text{(b)} \Bigg\} \tag{3.10}$$

may be found by finding y in terms of x in equation (3.10b) and substituting in equation (3.10a)

$$a_1x + \frac{b_1}{b_2}(c_2 - a_2x) = c_1 \tag{3.11}$$

rearranging gives

$$x = \frac{c_1b_2 - c_2b_1}{a_1b_2 - a_2b_1} \tag{3.12}$$

Providing $a_1b_2 - a_2b_1 \neq 0$ a unique solution for x is available.

$a_1b_2 - a_2b_1$ may be rewritten as $\begin{vmatrix} a_1 & b_1 \\ a_2 & b_2 \end{vmatrix}$ and

is called a second order determinant (two rows and two columns).

Thus using this notation

$$x = \frac{\begin{vmatrix} c_1 & b_1 \\ c_2 & b_2 \end{vmatrix}}{\begin{vmatrix} a_1 & b_1 \\ a_2 & b_2 \end{vmatrix}} \qquad y = \frac{\begin{vmatrix} c_1 & a_1 \\ c_2 & a_2 \end{vmatrix}}{\begin{vmatrix} a_1 & b_1 \\ a_2 & b_2 \end{vmatrix}} \tag{3.13}$$

or

$$\frac{x}{\begin{vmatrix} c_1 & b_1 \\ c_2 & b_2 \end{vmatrix}} = \frac{y}{\begin{vmatrix} c_1 & a_1 \\ c_2 & a_2 \end{vmatrix}} = \frac{1}{\begin{vmatrix} a_1 & b_1 \\ a_2 & b_2 \end{vmatrix}} \qquad (3.14)$$

A system of three equations in three unknowns

$$a_1 x + b_1 y + c_1 z = d_1$$
$$a_2 x + b_2 y + c_2 z = d_2 \qquad (3.15)$$
$$a_3 x + b_3 y + c_3 z = d_3$$

has the solution

$$\frac{x}{\begin{vmatrix} b_1 & c_1 & d_1 \\ b_2 & c_2 & d_2 \\ b_3 & c_3 & d_3 \end{vmatrix}} = \frac{y}{\begin{vmatrix} a_1 & c_1 & d_1 \\ a_2 & c_2 & d_2 \\ a_3 & c_3 & d_3 \end{vmatrix}} = \frac{z}{\begin{vmatrix} a_1 & b_1 & d_1 \\ a_2 & b_2 & d_2 \\ a_3 & b_3 & d_3 \end{vmatrix}} = \frac{1}{\begin{vmatrix} a_1 & b_1 & c_1 \\ a_2 & b_2 & c_2 \\ a_3 & b_3 & c_3 \end{vmatrix}} \qquad (3.16)$$

where

$$\begin{vmatrix} a_1 & b_1 & c_1 \\ a_2 & b_2 & c_2 \\ a_3 & b_3 & c_3 \end{vmatrix}$$

is defined to mean

$$a_1 \begin{vmatrix} b_2 & c_2 \\ b_3 & c_3 \end{vmatrix} - a_2 \begin{vmatrix} b_1 & c_1 \\ b_3 & c_3 \end{vmatrix} + a_3 \begin{vmatrix} b_1 & c_1 \\ b_2 & c_2 \end{vmatrix}$$

or

$$a_1(b_2 c_3 - c_2 b_3) - a_2(b_1 c_3 - c_1 b_3) + a_3(b_1 c_2 - c_1 b_2)$$

The other determinants may be evaluated in a similar manner.

Note in equation (3.16) that the determinant of the last fraction is obtained from writing the coefficients in the form that they appear on the left-hand side of the original equations (3.15). This is a determinant of the third order.

The leading element of the determinant is a_1. If the row and column containing the element is omitted then the determinant of the remaining elements is called the Minor of that element.

e.g. the Minor of a_3 is $\begin{vmatrix} b_1 & c_1 \\ b_2 & c_2 \end{vmatrix}$

As a general rule, higher orders of determinants can be evaluated by multiplying the elements in the first column by their Minors. Starting with the leading element attach + and − signs alternatively to these products.

3.2.3 Cofactors

The + and − signs associated with the Minors may be set out as shown below

$$
\begin{vmatrix}
+ & - & + & - & \cdots \\
- & + & - & + & \cdots \\
+ & - & + & - & \cdots \\
- & + & - & + & \cdots \\
\vdots & \vdots & \vdots & \vdots &
\end{vmatrix}
$$

Fig. 3.2

If the signs in Fig. 3.2 are attached to the corresponding Minors the cofactors of the elements are obtained denoted by the capital letter corresponding to the element.

$$
\begin{vmatrix}
a_1 & b_1 & c_1 \\
a_2 & b_2 & c_2 \\
a_3 & b_3 & c_3
\end{vmatrix}
$$

$$
A_2 = - \begin{vmatrix} b_1 & c_1 \\ b_3 & c_3 \end{vmatrix} \qquad
B_2 = \begin{vmatrix} a_1 & c_1 \\ a_3 & c_3 \end{vmatrix} \qquad
C_2 = - \begin{vmatrix} a_1 & b_1 \\ a_3 & b_3 \end{vmatrix}
$$

Determinants may be evaluated by expanding along any row or column by multiplying each element in the row or by its cofactor and adding the products.

Referring again to equation (3.16) it can be seen that the solution for x given by

$$
x = \frac{\begin{vmatrix} b_1 & c_1 & d_1 \\ b_2 & c_2 & d_2 \\ b_3 & c_3 & d_3 \end{vmatrix}}{\begin{vmatrix} a_1 & b_1 & c_1 \\ a_2 & b_2 & c_2 \\ a_3 & b_3 & c_3 \end{vmatrix}} \tag{3.17}
$$

is based purely on the coefficients of the original system equation (3.15). It is therefore possible, when using computers, for the solution to simultaneous equations to store and manipulate the coefficients separately from the variables. It is also useful to use matrix notation to write the equations in shorthand form.

The general system of n variables and n unknowns may be written as

$$\left.\begin{array}{c} a_{11}\,x_1 + a_{12}\,x_2 + a_{13}\,x_3 - - - a_{1n}\,x_n = b_1 \\ \\ a_{n1}\,x_1 + a_{1n}\,x_2 + a_{13}\,x_3 - - - a_{nn}\,x_n = b_n \end{array}\right\} \qquad (3.18)$$

The coefficients of the variables may be written as a matrix

$$A = \begin{bmatrix} a_{11} & a_{12} - - - a_{13} \\ & a_{ij} \\ a_{n1} & a_{n2} - - - a_{nn} \end{bmatrix} \qquad \leftarrow \text{row } i \qquad (3.19)$$

$$\uparrow$$
$$\text{Column } j$$

and the coefficient a_{ij} denotes the coefficient of x_j in the ith equation.
The constants b_i may be written as a column vector

$$b = \begin{bmatrix} b_1 \\ b_2 \\ b_3 \\ \vdots \\ b_n \end{bmatrix}$$

hence the system equation (3.18) may be written as

$$Ax = b \qquad (3.19)$$

where A is a matrix and x and b are column vectors. Equation (3.19) may be solved by writing

$$x = A^{-1}b \qquad (3.20)$$

A^{-1} is the inverse of A and from matrix algebra

$$A^{-1}A = I$$

where I is the Identity matrix with one's on the principal diagonal and zeros elsewhere.

$$I = \begin{bmatrix} 1 & 0 & 0 - - - 0 \\ 0 & 1 & 0 & 0 \\ 0 & 0 & 1 & 0 \\ \vdots & & & \vdots \\ 0 & 0 & 0 - - - 1 \end{bmatrix} \qquad (3.21)$$

A^{-1} may be defined as

$$A^{-1}\left[\frac{C}{|A|}\right] \qquad (3.22)$$

where

(1) C is the adjoint of A and is the transpose of the matrix of cofactors (i.e. the rows and columns have been interchanged).

$$C = \begin{bmatrix} A_{11} & A_{21} & A_{31} \\ A_{12} & A_{22} & A_{32} \\ A_{13} & A_{23} & A_{33} \end{bmatrix}$$

(2) $|A|$ is the determinant of A.

(3) $A_{11} \rightarrow A_{33}$ are the cofactors of A.

Hence

$$x = \left[\frac{C}{|A|} \right] \times b \tag{3.23}$$

Taking a numerical example by writing (4) (5) and (6) in the form of equation (3.19)

$$x_1 + x_2 + x_3 = 9$$
$$2x_1 - 3x_2 + x_3 = 1 \tag{3.24}$$
$$x_1 - x_2 + 3x_3 = 7$$

$$A = \begin{bmatrix} 1 & 1 & 1 \\ 2 & -3 & 1 \\ 1 & -1 & 3 \end{bmatrix} \qquad b = \begin{bmatrix} 9 \\ 1 \\ 7 \end{bmatrix} \qquad C = \begin{bmatrix} A_{11} & A_{21} & A_{31} \\ A_{12} & A_{22} & A_{32} \\ A_{13} & A_{23} & A_{33} \end{bmatrix}$$

$$|A| = 1 \times (-3 \times 3 - 1 \times -1) - 1 \times (2 \times 3 - 1 \times 1)$$
$$+ 1 \times (2 \times -1 - -3 \times 1) = -12$$

$$A_{11} = + \begin{vmatrix} -3 & 1 \\ -1 & 3 \end{vmatrix} = -8$$

$$A_{12} = - \begin{vmatrix} 2 & 1 \\ 1 & 3 \end{vmatrix} = -5$$

$$A_{13} = + \begin{vmatrix} 2 & -3 \\ 1 & -1 \end{vmatrix} = 1$$

The other cofactors may be similarly evaluated to give

$$C = \begin{bmatrix} -8 & -4 & 4 \\ -5 & 2 & 1 \\ 1 & 2 & -5 \end{bmatrix}$$

Thus

$$x = \frac{\begin{bmatrix} -8 & -4 & 4 \\ -5 & 2 & 1 \\ 1 & 2 & -5 \end{bmatrix} \begin{bmatrix} 9 \\ 1 \\ 7 \end{bmatrix}}{-12} \tag{3.25}$$

x_1, x_2 and x_3 may be evaluated using the rule for matrix multiplication, i.e. the element in each row of the first matrix is multiplied by the corresponding element in the column of the second matrix

$$x_1 = (-8 \times 9 + -4 \times 1 + 4 \times 7)/-12 = +4$$

$$x_2 = (-5 \times 9 + 2 \times 1 + 1 \times 7)/-12 \quad = +3 \tag{3.26}$$

$$x_3 = (1 \times 9 + 2 \times 1 - 5 \times 7)/-12 \quad = +2$$

This method of solving equations is known as Cramers' rule.

It is not generally considered useful as a numerical technique because it involves the calculation of n^2 determinants each having order $n-1$.

There are many methods available for solving systems of linear equations— most algorithms are designed in some way to minimize computer time or allocate storage space efficiently for large systems in which many of the coefficients have zero value.

Two methods will be described here to illustrate the concepts on which more sophisticated packages are based.

3.2.4 Gauss elimination method

Using matrix algebra it is possible to transform the system of equations into an equivalent system of triangular form which can be solved by a back substitution process.

$$x_1 + x_2 + x_3 = 9 \quad \text{(A)}$$

$$2x_1 - 3x_2 + x_3 = 1 \quad \text{(B)} \tag{3.27}$$

$$x_1 - x_2 + 3x_3 = 7 \quad \text{(C)}$$

The first step is to eliminate x_1 from equations (3.27) (B) and (C)

$$x_1 + x_2 + x_3 = 9 \qquad \text{(A}_1\text{)}$$

$$0 - 2.5x_2 - 0.5x_3 = -8.5 \quad \text{(B}_1\text{)} \quad ((\text{B})/2 - (\text{A}))$$

$$0 - 2x_2 + 2x_3 = -2 \qquad \text{(C}_1\text{)} \quad ((\text{C}) - (\text{A}))$$

The second step is to eliminate x_2 from equation (3.27) (C_1)

$$x_1 + x_2 + x_3 = 9 \qquad (A_2)$$

$$0 - 2.5x_2 - 0.5x_3 = -8.5 \quad (B_2) \quad ((C_1/4 + (B_2)))$$

$$0 + 0 + 12x_3 = 24 \qquad (C_2) \quad ((5x(C_1) - 4x(B_1)))$$

The coefficient matrix is now

$$\begin{bmatrix} 1 & 1 & 1 \\ 0 & 4 & 1 \\ 0 & 0 & 5 \end{bmatrix}$$

which is an upper triangular matrix with all coefficients zero below the principal diagonal.

From (C_2) $x_3 = 2$

From (B_2) $-2.5x_2 - 1 = 7.5$ $x_2 = \dfrac{7.5}{2.5} = 3$

From (A_2) $x_1 = 3 + 2 = 9$ $x_1 = 4$

In general system equation (3.27) may be expressed as

$$a_{11}x_1 + a_{12}x_2 + a_{13}x_3 = b_1 \quad (a)$$

$$a_{21}x_1 + a_{22}x_2 + a_{23}x_3 = b_2 \quad (b) \qquad (3.28)$$

$$a_{31}x_1 + a_{32}x_2 + a_{33}x_3 = b_3 \quad (c)$$

which may be written in augmented matrix form as

$$\begin{bmatrix} a_{11} & a_{12} & a_{13} & b_1 \\ a_{21} & a_{22} & a_{23} & b_2 \\ a_{31} & a_{32} & a_{33} & b_3 \end{bmatrix} \begin{matrix} (a) \\ (b) \\ (c) \end{matrix} \qquad (3.29)$$

a_{21} may be eliminated by

    (1) specifying a multiplier $-\dfrac{a_{21}}{a_{11}}$

    (2) adding $-\dfrac{a_{21}}{a_{11}} \times$ row (a) (equation (3.29)) to row (b) (equation (3.29))

a_{31} may be eliminated by

    (1) specifying multiplier $-\dfrac{a_{31}}{a_{11}}$

    (2) adding $-\dfrac{a_{31}}{a_{11}} \times$ row (a) (equation (3.29) to row (b) (equation (3.29)).

Hence a new matrix is formed

$$\begin{bmatrix} a_{11} & a_{12} & a_{13} & b_1 \\ 0 & a_{22}' & a_{23}' & b_2' \\ 0 & a_{32}' & a_{33}' & b_3' \end{bmatrix} \begin{matrix} \text{(a1)} \\ \text{(b1)} \\ \text{(c1)} \end{matrix} \qquad (3.30)$$

The values of elements in row (a1) (equation (3.30)) remain the same elements in rows (b1) (equation (3.30)) (c1) (equation (3.30)) such as a_{22}' have a new value obtained from $a_{22} - a_{12}(a_{21}/a_{11})$ etc.

The element a_{32}' may be eliminated by

(1) Specifying a multiplier $-\dfrac{a_{32}'}{a_{22}'}$

(2) adding $-(a_{32}'/a_{22}') \times$ row (b1) (equation (3.30)) to row (c1) (equation (3.30)).

The final system of equations is now

$$a_{11}x_1 + a_{12}x_2 + a_{13}x_3 = b_1$$

$$0 + a_{22}'x_2 + a_{23}'x_3 = b_2' \qquad (3.31)$$

$$0 + 0 + a_{33}''x_3 = b_3$$

which may be solved by back substitution

$$x_3 = \frac{b_3}{a_{33}''}$$

$$x_2 = (b_2' - a_{23}'x_3)/a_{22}' \qquad (3.32)$$

$$x_1 = (b_1 - a_{13}x_3 - a_{12}x_2/a_{11})$$

The diagonal elements a_{11}, a_{22}' and a_{33}'' are known as pivot elements and occur in the denominator of the multipliers and also in the back substitution process. To carry out the successive elimination procedure it is necessary for the pivot elements to be non-zero. If at any stage the pivot element is zero the remaining rows may be interchanged to produce a non-zero pivot. If it is not possible to find a non-zero pivot the system of linear equations has no solution.

It is not possible here to discuss in great detail the accuracy of a numerical solution to a system of linear equations. However, some of the sources of error are indicated below.

(1) The value of the pivot elements can affect the answer significantly. If the pivot element is small compared to other elements in its column which have to be eliminated then the multiplier will be greater than one in magnitude which may lead to an increase in round-off errors. It is possible to minimize this problem by arranging, at each elimination stage,

the rows of the matrix below the pivot so that the new pivot is larger in absolute value than any element beneath it in its column. Thus multipliers have a magnitude less than or equal to one.

(2) Uncertainty in the coefficients of the variables may also lead to errors particularly if the coefficients are obtained from experimental observation. This problem may be even more severe if the equations are ill-conditioned in which case the solutions are very sensitive to small changes in coefficient value.

(3) Round-off errors may be significant particularly in large systems of equations. These errors are propogated at each step in the solution procedure. The growth of such errors can lead to completely useless results.

3.2.5 An iterative method

Large systems of linear equations cannot be solved with any degree of certainty by the direct approach described above. Several iterative methods have been developed one of which is the Gauss–Seidel approach. As with all iterative methods, it starts with an initial approximate solution and uses this in a recurrence formula to provide another approximation. Several repetitions produce a sequence of solutions which under suitable conditions converge to the exact solution.

Consider again equation (3.27) which may be rewritten as

$$x_1 = 9 - x_2 + x_3 \qquad \text{(a)}$$

$$x_2 = (-1 + 2x_1 + x_3)/3 \quad \text{(b)} \qquad (3.33)$$

$$x_3 = (7 - x_1 + x_2)/3 \qquad \text{(c)}$$

Assuming an initial solution of $x_1 = x_2 = x_3 = 0$ after 16 iterations

$$x_1 = 4.00469$$

$$x_2 = 3.00337$$

$$x_3 = 1.99956$$

The exact solution for these equations found previously was

$$x_1 = 4$$

$$x_2 = 3$$

$$x_3 = 2$$

The number of iterations used here was quite arbitrary. In practice, the iterations would be stopped using criteria based on the relative values at successive iterative steps. One such rule is to end when the modulus of the difference between the sum of the x values at the nth step and the $(n + 1)$th step are less than a given value e.g.

$$\sum_i |x_i^{n+1} - x_i^n| < 0.00001$$

It is difficult to estimate the number of iterations required to achieve a particular accuracy. Convergence may in some cases be very quick and in other cases be too slow to be useful. It is possible to increase the chances of satisfactory convergence by rearranging the equations where possible so that the leading diagonal coefficient has an absolute value greater than others in its row. If the equations are badly arranged, it is possible for this method to produce results which diverge from the correct answer.

A further example of a numerical technique for the solution of a system of linear equations which may be written in tri-diagonal matrix form, is given in Chapter 6.

3.3 FINITE DIFFERENCE APPROXIMATIONS TO DIFFERENTIAL EQUATIONS

Before discussing finite difference methods it is useful to consider the way in which Taylor's series may be used to evaluate a function using a known value of the function and its derivatives close to the point of interest.

If it is assumed that the value of y at point B on the curve is $f(t)$ and that point C has a value of $f(t + h)$, then the value at C may be expressed in terms of the value at B in the following way.

$$f(t + h) = f(t) + h \cdot \frac{\mathrm{d}f(t)}{\mathrm{d}t} + \frac{h^2}{2!} \cdot \frac{\mathrm{d}^2 f(t)}{\mathrm{d}t^2} + \frac{h^3}{3!} \cdot \frac{\mathrm{d}^3 f(t)}{\mathrm{d}t^3} \qquad (3.34)$$

Similarly point A can be evaluated from point B

$$f(t - h) = f(t) - h \frac{\mathrm{d}f(t)}{\mathrm{d}t} + \frac{h^2}{2!} \cdot \frac{\mathrm{d}^2 f(t)}{\mathrm{d}t^2} + \frac{h^3}{3!} \frac{\mathrm{d}^3 f(t)}{\mathrm{d}t^3} \qquad (3.35)$$

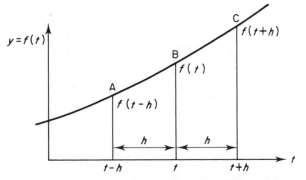

Fig. 3.3 Taylor's series approximation of a differential equation

Assuming that $f(t)$ is continuously differentiable at B, the values calculated at C may be calculated to any degree of accuracy depending on the number of terms involved in the Taylor series. Although Taylor series are of no practical use for solving differential equations they are a useful basis for evaluating and comparing alternative numerical solution methods. For example, if a solution method agrees with the first three terms of the Taylor series then the truncation error is of order h^3, i.e. terms including the third derivative and all higher derivatives are ignored.

Consider the first-order differential equation

$$\frac{dy}{dt} = ky \qquad (3.36)$$

which has a theoretical solution

$$y = A\, e^{kt} \qquad (3.37)$$

where A is an arbitrary constant.

If A is assumed equal to one then in Fig. 3.4 where $t = 0$, $y = 1$. Equation (3.35) may be used to evaluate y for any given value of t. Unfortunately, most differential equations are not amenable to analytic solution and may best be tackled by a numerical approach.

Equation (3.36) may be solved numerically given the initial value of $y = 1$ at $t = 0$. The numerical solution involves estimating successive values of y at points t_1, t_2, $t_3 \ldots t_n$. The spacing of t_1, t_2, etc. is usually equidistant.

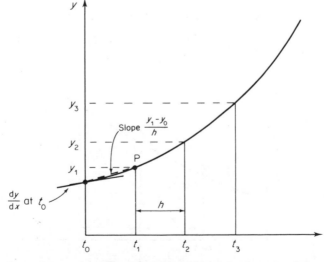

Fig. 3.4 Truncation error

The Taylor series gives one method of obtaining a numerical solution. From equation (3.34), the first two terms of the expension may be written as

$$F(t+h) \simeq f(t) + h \cdot \frac{\mathrm{d}f(t)}{\mathrm{d}x} \tag{3.38}$$

rearranging gives

$$\frac{\mathrm{d}}{\mathrm{d}t} f(t) = \frac{f(t+h) - f(t)}{h} \tag{3.39}$$

From Fig. 3.3

$$f(t) = y_0 = 1 \text{ evaluated at } t = 0 = t_0$$

$f(t+h)$ is the value of y_1 at point P evaluated at t_1.
   Also

$$\frac{\mathrm{d}f(t)}{\mathrm{d}t} = ky_0 \quad \text{from equation (3.36)}$$

Thus equation (3.39) may be rewritten as

$$\frac{y_1 - y_0}{h} = ky_0$$

or

$$y_1 = y_0 + h \cdot ky_0 \tag{3.40}$$

This is known as Euler's method in which the derivative evaluated at the point y_0, t_0 is used to approximate the true value of y_1 as shown in Fig. 3.5.
   Having estimated y_1 this may be used to calculate y_2

$$\frac{y_2 - y_1}{h} = ky_1$$

$$y_2 = y_1 + hky_1 \tag{3.41}$$

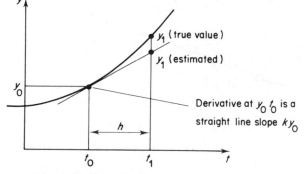

Fig. 3.5   Simple Euler method for solving differential equations

and in general

$$y_{n+1} = y_n + hky_n \qquad (3.42)$$

Euler's method has truncation errors of order h^2 and unless h is kept very small the errors become large and the results inaccurate.

3.3.1 Runge–Kutta methods

The Runge–Kutta approach is based on the same principles as the Euler method but improves the accuracy of the result by reducing the truncation error when compared term by term with the Taylor expansion.

Consider again equation (3.36) which may be rewritten in the more general form

$$\frac{dy}{dx} = f(y) \qquad (3.43)$$

In words, the slope of the derivative is equal to $f(y)$ and in Fig. 3.5 ky_0 may be represented by $f(y_0)$ (the derivative evaluated at y_0, t_0).

The Runge–Kutta approach uses the derivative evaluated at more than one point to estimate the new value of y_1.

A second-order method uses the average of slopes calculated at t_0, y_0, and at the point t_1, y_{1e} (the first estimate of y_1) to improve the approximation. This is shown graphically in Fig. 3.4.

Now

$$\frac{dy}{dt} \simeq \frac{y_1 - y_0}{h} = \tfrac{1}{2}(f(y_0) + f(y_{1e})) \qquad (3.44)$$

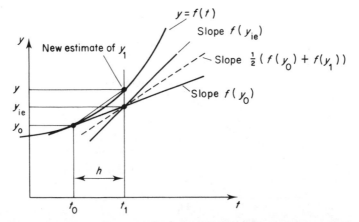

Fig. 3.6 Runge–Kutta method for solving differential equations

and

$$y_1 = y_0 + \frac{h}{2} f(y_0) + \frac{h}{2} f(y_{1e}) \tag{3.45}$$

often equation (3.45) is expressed in the form

$$y_1 = y_0 + \tfrac{1}{2}(k_1 + k_2) \tag{3.46}$$

where

$$k_1 = hf(y_0)$$

$$k_2 = hf(y_{1e})$$

now

$$y_{1e} = y_0 + hf(y_0) = y_0 + k_1 \tag{3.47}$$

and

$$k_2 \perp hf(y_0 + k_1)$$

For a general function with the derivative a function of both y and t

$$\frac{dy}{dt} = f(t, y) \tag{3.48}$$

The second-order Runge–Kutta can be written as

$$k_1 = h \cdot f(t_n, y_n)$$

$$k_2 = h \cdot f(t_n + h, y_n + k_1)$$

$$y_{n+1} = y_n + \tfrac{1}{2}(k_1 + k_2) \tag{3.49}$$

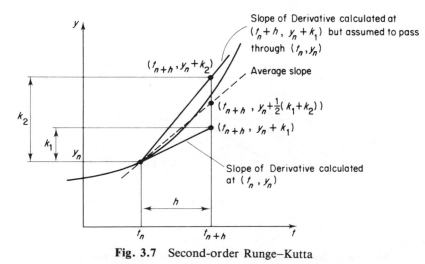

Fig. 3.7   Second-order Runge–Kutta

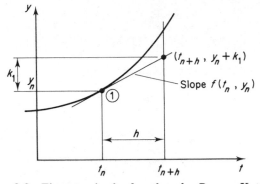

Fig. 3.8 First step in the fourth-order Runge–Kutta

Fig. 3.7 shows graphically the new value of y_{n+1} based on the average of the slopes calculated at (t_n, y_n) and $(t_n + h, y_n + k_1)$

A common method in use is the fourth-order method in which the average slope used to calculate $y_n + 1$ is based on a weighted average of the derivative evaluated at four separate points within the increment $t_n, t_n + h$.

The fourth-order method for $dy/dt = f(t, y)$ may be written as

$$y_{n+1} = y_n + \tfrac{1}{6}(k_1 + 2k_2 + 2k_3 + k_4) \tag{3.50}$$

where

$$k_1 = hf(t_n, y_n)$$
$$k_2 = hg(t_n + \tfrac{1}{2}h, y_n + \tfrac{1}{2}k_1)$$
$$k_3 = hf(t_n + \tfrac{1}{2}h, y_n + \tfrac{1}{2}k_2)$$
$$k_4 = hg(t_n + h, y_n + k_3)$$

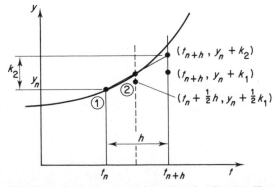

Fig. 3.9 Second step in fourth-order Runge–Kutta

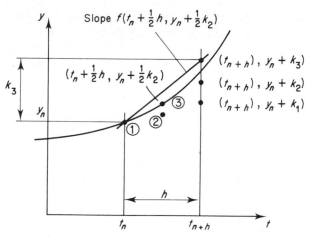

Fig. 3.10 Third step in fourth-order Runge–Kutta

A graphical explanation of the procedure to evaluate $k_1 \ldots k_4$ and hence obtain y_1 is shown below.

Using the derivative evaluated at point 1, the value of k_1 can be calculated from $h \cdot f(t_n, y_n)$. k_1 is the increment of y corresponding to the increment h and the estimate of slope $f(t_n, y_n)$.

Point 2 has coordinates $(t_n + h/2, y_n + (k_1/2)$ and the slope of the derivative of the function at point 2 is $f(t_n + h/2, y_n + (k_1/2)$. k_2 is the increment of y corresponding to the increment h and estimate of slope $f(t_n + h/2, y_n + (k_1/2))$ assuming that the slope passes through point 1.

Point 3 has coordinates $(t_n + h/2, y_n + \frac{1}{2}k_2)$ and the slope of the derivative of the function at point 3 is $f(t_n + h/2, y_n + (k_2/2))$. k_3 is the increment in y

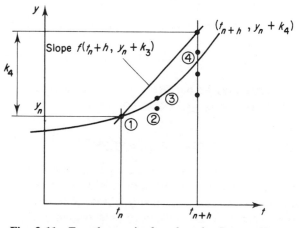

Fig. 3.11 Fourth step in fourth-order Runge–Kutta

corresponding to the increment h and the estimate of the slope $f(t_n + (h/2)$, $y_n + (k_2/2))$ assuming that the slope passes though point 1.

Point 4 has coordinates $(t_n + h, y_n + k_3)$ and the slope of the derivative of the function at point 4 is $f(t_n + h)$, $(y_n + k_3)$. k_4 is the increment in y corresponding to the increment h and the estimate of the slope $f(t_n + h)$, $y_n + k_3$ assuming that the slope passes through point 1.

Thus four approximations to the derivative dy/dt at t_n, y_n have been made producing four estimates for $y_n + 1$. These are combined as a weighted average as shown in equation (3.50).

It is again worth repeating that the difference approach to evaluating a function, by using incremental steps in the dependant variable, is an iterative procedure. Having evaluated $y_n + 1$ for one time step, this is then used as y_n in the next time step.

Runge–Kutta methods are classified as single-step methods because the only value of the approximate solution used in constructing $y_n + 1$ is y_n. Only one starting value of the function is needed to being the calculation using this method.

3.3.2 Multi-step methods

Multi-step methods make use of earlier values like y_{n-1}, y_{n-2} etc. in order to reduce the number of times y has to be evaluated. Several methods are available one of which is Milnes fourth-order method

$$\frac{dy}{dx} = f(x, y)$$

$$y_{n+1} = y_{n-3} + \frac{4h}{3}\left(2f(x_n, y_n) - f(x_{n-1}, y_{n-1}) + 2f(x_{n-2}, y_{n-2})\right) \quad (3.51)$$

More than one value of the function is required to start the calculation procedure which is a disadvantage however to construct y_{n+1} $f(xy)$ need only be evaluated once since $f(x_{n-1}, y_{n-1})$, $f(x_{n-2}, y_{n-2})$ etc. have already been calculated.

Often Runge–Kutta may be used to start the calculation and then a multi-step approach used to speed up further computation.

3.4 PARTIAL DIFFERENTIAL EQUATIONS

Many partial differential equations describing pollution dispersion and decay in the environment cannot be solved analytically because of variable coefficients, boundary conditions or complexity. Finite difference methods are frequently used even when an analytical solution is available. Difference methods are approximate in the sense that an instantaneous derivative is approximated by a difference quotient over a small interval. However, they are not approximate in the sense of being crude estimates. Data used in environmental problems are subject to errors in measurement and also arithmetical calculations are subject to round-off errors so the even analytical solutions give approximate numerical answers. Finite difference methods give solutions as accurate as the data warrants or as accurately as is necessary for the technical purposes for which the solution is required. In both cases the finite difference solution is as satisfactory as one calculated from an analytical formula.

The method may be illustrated using the one dimensional convective diffusion equation for a non-conservative substance with a first-order decay rate.

$$\frac{\partial c}{\partial t} = E \frac{\partial^2 c}{\partial x^2} - U \frac{\partial c}{\partial x} - kc + La \qquad (3.52)$$

Equation (3.52) may be used to describe an inland river with unidirectional flow, constant dispersion coefficient, constant cross-sectional area and constant velocity. Pollutant concentration will vary both with distance along the river and with time.

In the finite difference approach the area of integration is overlayed by a rectangular mesh and the approximate solution to the differential equation is found at the points of intersection of the mesh. In Fig. 3.12, the x-axis is divided into increments of h and the time axis in increments of k. $C_{i,n}$ is the concentration value at the end of the ith distance interval and the nth time interval.

The partial derivatives in equation (3.52) are approximated using values at

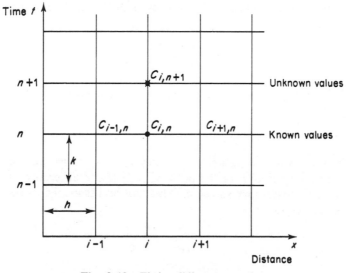

Fig. 3.12 Finite difference mesh

neighbouring mesh points to produce algebraic equations which may be solved directly or simultaneously depending on the form of the difference method employed.

Explicit methods

The process is similar to that for ordinary differential equations in which the solution is obtained by stepping through time starting with a known concentration distribution. In the Explicit method the Unknown value $C_{i,n+1}$ is obtained using the three known values $C_{i-1,n}$, $C_{i,n}$ and $C_{i+1,n}$ to approximate the derivatives.

Examining each term separately:

1. The zero-order term kC can be represented by

$$kC = K \frac{(C_{i,n} + C_{i,n+1})}{2} \tag{3.53}$$

and

$$L_a = \Delta L_i = \text{the increment of load discharged during the time step at the point } i.$$

2. The first-order term $\partial c/\partial t$ must be represented by a forward difference approximation in order to advance the time step of the computation.

$$\frac{\partial c}{\partial t} = \frac{C_{i,n+1} - C_{i,n}}{k} \tag{3.54}$$

The term $U(\partial c/\partial x)$ may be represented by a backward difference approximation

$$U \frac{\partial c}{\partial x} = \frac{C_{i,n} - C_{i,-1,n}}{h} \tag{3.55}$$

This is necessary because any method which represents pure advection requires that

$$C_{i,n+1} = C_{i-1,n} \tag{3.56}$$

In other words, a particle at the point $C_{i-1,n}$ must move a distance h in the time interval k to the point $C_{i,n+1}$ if the difference method is to provide a satisfactory solution. This can be illustrated using the simple advection equation

$$\frac{\partial c}{\partial x} = -U \frac{\partial c}{\partial x}$$

$$\frac{c_{i,n+1} - c_{i,n}}{k} = -U \frac{(c_{i,n} - c_{i-1,n})}{h} \tag{3.57}$$

If the size of the mesh increments are chosen correctly so that

$$\frac{h}{K} = U$$

then it can be seen from equation (3.57) that the requirement specified by equation (3.56) is satisfied.

It can be shown that a forward difference or central difference approximation cannot satisfy this condition. Intuitively it can be seen that under conditions of pure convection the concentration at point i cannot depend on the concentration at $i + 1$.

3. The dispersion term $E(\partial^2 c/\partial x^2)$ may be represented by a central difference approximation.

$$E\frac{\partial^2 c}{\partial x^2} = E\frac{(c_{i+1,n} - 2c_{i,n} + c_{i-1,n})}{h^2} \tag{3.58}$$

This is derived from consideration of a Taylor series expansion about the point x_i. Taking the first three terms only for $c(x_i + h)$ and $c(x_i - h)$

$$c(x_i + h) = c(x_i) + \frac{h}{dx}\frac{dc(x_i)}{dx} + \frac{h^2}{2}\frac{d^2 c(x_i)}{dx^2}$$

$$c(x_i - h) = c(x_i) - \frac{h}{dx}\frac{dc(x_i)}{dx} + \frac{h^2}{2}\frac{d^2 c(x_i)}{dx^2}$$

Adding the above gives:

$$c(x_i + h) + c(x_i - h) = 2c(x_i) + h^2\frac{d^2 c(x_i)}{dx^2} \tag{3.59}$$

rearranging equation (3.59) gives

$$\frac{d^2 c(x_i)}{dx^2} = \frac{1}{h^2}[c(x_i + h) - 2c(x_i) + c(x_i - h)] \tag{3.60}$$

which is in the same form as equation (3.58) if $c(x_2 + h)$ is written as c_{n+1} etc.

It can be shown that a necessary condition for stability of solutions with second derivatives of this form is

$$\frac{kE}{h^2} \leqslant \frac{1}{2} \tag{3.61}$$

The finite difference approximation to equation (3.52) can now be written as

$$\frac{c_{i,n+1} - c_{i,n}}{k} = E\frac{(c_{i+1,n} - 2c_{i,n} + c_{i-1,n})}{h^2}$$

$$- U\frac{(c_{i,n} - c_{i-1,n})}{h} - k\frac{(c_{i,n} + c_{i,n+1})}{2} + \Delta La \tag{3.62}$$

Equation (3.62) may be rearranged to solve for the unknown value $c_{i,n+1}$ in terms of the known values of c at the nth time level. By solving equation (3.62) for each x mesh point at the $n + 1$th time level the new concentration profile can be computed. It is generally necessary and convenient to choose the spatial boundaries of the model in such a way that boundary concentrations are specified as constant for all time intervals.

It has been shown that the direct use of equation (3.62) can produce initial computational errors which may give misleading results. A two-step method has been suggested to overcome these problems. In this procedure the load of pollution is first convected downstream for one time step and then other processes such as dispersion and decay are carried out. In Fig. 3.13 the concentrations are moved one mesh to the right starting at the right-hand end, i.e.

$$c_{i+3,n} = k \text{ Boundary conditions}$$

$$c_{i+2,n} = c_{i+1,n}$$

$$c_{i+1,n} = c_{i,n}$$

$$c_{i,n} = c_{1-1,n}$$

$$c_{i-1,n} = c_{i-1,n}$$

$$c_{i-2,n} = c_{1-2,n+1} \text{ Boundary condition}$$

Step 1

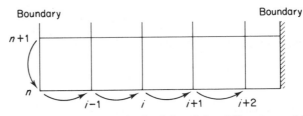

Fig. 3.13 Explicit method of solving finite difference problem

Step 2

The new values of $c_{i,n}$ etc. can be used in equation (3.62) which no longer contains the advective term $U(dc/dx)$.

Implicit methods

Implicit methods have been developed to overcome the restrictions of the explicit method, such as the ratio of the time and distance mesh increments.

In the implicit method, six points are used to approximate the derivatives at the i mesh distance leading to an equation involving three unknowns in the

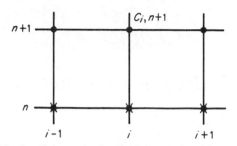

Fig. 3.14 Implicit method of solving finite difference problems

form:

$$\alpha c_{i-1, n+1} + \beta c_{i, n+1} + \gamma c_{i+1, n+1} = \delta(c_{i-1, n}, c_{1, n}, c_{i+1, n})$$

Equations are evaluated at each mesh point in the x direction and solved simultaneously at each time step. It is necessary for the boundary conditions to be specified so that the end equations only have two unknowns. A more comprehensive description of this approach is given in Chapter 6 on Estuary models.

3.5 BIBLIOGRAPHY

Dorn and McCracken (1972). *Numerical Methods with Fortran IV Case Studies*, John Wiley & Sons, New York.

Hosking, R. J., Joyce, D. C., and Turner, J. C. (1978). *First Steps in Numerical Analysis*, Hodder & Stoughton, London.

Smith, J. M. (1977). *Mathematical Modelling and Digital Simulation for Engineers and Scientists*, John Wiley & Sons, New York.

Froberg, C. E. (1970). *Introduction to Numerical Analysis*, (2nd Edition). Addison-Wesley Publishing Co., New York.

Smith, G. D. (0000). *Numerical Solution of Partial Differential Equations: Finite Difference Methods*, Clarendon Press, Oxford.

Dresnack, R. and Dobbins, W. E. (1968) Numerical Analysis of BOD and DO profiles, Proc. ASCE., SA5 pp. 789–807

Chapter 4

Modelling of Kinetics

A. JAMES

4.1 INTRODUCTION

Many water-quality models attempt to simulate changes in the concentration of substances which are in true solution, colloidal solution or in suspension. There are some substances which are sufficiently inert for their concentration to be regarded as unchanging (except by dilution). These are referred to as conservative substances and are often used as tracers.

However, the majority of substances in water are subject to change in concentration due to physical, chemical and biological processes. An understanding of these processes is therefore useful in constructing water-quality models of non-conservative substances. Also many of the processes employed in the treatment of water and waste-waters rely on biological agencies (mainly bacteria) so an understanding of the kinetics of growth is helpful in design and operation.

The following notes describe the ways of representing physical, chemical and biological kinetics. Examples of these can be found throughout Chapters 5–14 which deal with the applications of modelling to natural waters and wastewater treatment.

4.2 THE KINETICS OF CHEMICAL AND PHYSICAL PROCESSES

There is no clear dividing line between physical and chemical phenomena so all non-biological processes are considered under this heading.

Only those processes which commonly occur in water-quality models have been included, due to limitations of space but the same basic principles of simulation may be used for other phenomena like thermodynamics and the basic equations may be obtained from standard texts on physical chemistry.

4.2.1 Solution equilibrium

Many chemical reactions taking place in solution are reversible to some extent and reaction takes place until a state of balance or chemical equilibrium is reached for the opposing processes.

For the situation in which materials A and B react to form products C and

D the equilibrium equation can be written as

$$mA + nB \rightleftharpoons pC + qD \qquad (4.1)$$

where

A, B, C, D are molecules or ions

m, n, p, q are coefficients used to balance the equation.

From the Law of Mass Action it can be shown that

$$\frac{[a_C]^p \ [a_D]^q}{[a_A]^m \ [a_B]^n} = K \qquad (4.2)$$

where

a_C etc. are activities in mg 1^{-1}. (Activities are related to concentration by $a = \gamma c$)

$\gamma =$ the activity coefficient, and

$c =$ concentration

$K =$ activity equilibrium constant characteristic of the particular equilibrium varying only with the temperature of the solution.

γ is approximately equal to unity for most solutions of non-electrolytes and also for dilute solutions of electrolytes so that concentrations can be used instead of activities in equation (4.2)

$$\frac{[C]^p \ [D]^q}{[A]^m \ [B]^n} = K \qquad (4.3)$$

and equation (4.3) is the one generally used in water-quality models for simulating reversible reactions (see Chapter 13 for an example).

4.2.2 Chemical kinetics

Equilibrium conditions as expressed by the Law of Mass Action do not always occur in the natural environment or in water and wastewater treatment processes. The speed of many reactions is slow relative to physical advective and dispersive processes. An understanding of the time dependence of the reaction is therefore often more important than knowledge of the final equilibrium condition.

The rate of reaction depends not only on the particular substances involved but also on the physical state of the system. Homogeneous systems are characterized by all reactants and products of the reaction occurring in the same physical state. Heterogeneous systems involve reactions between substances in two or more states and the rate of reaction may depend on factors other than the normal chemical reaction as expressed by the Law of Mass Action.

4.2.3 Homogeneous systems

Assuming a well-mixed homogeneous system, the manner in which the rate of reaction may vary with the concentration of some or all of the reacting substances is denoted by the order of the reaction. For the reaction

$$A + B \to P \tag{4.4}$$

The equation for the rate of formation of P can be written

$$\frac{d\ [P]}{dt} = k\ [A]^v\ [B]^w \tag{4.5}$$

where k is the rate constant (units t^{-1}) and A and B are instantaneous concentrations. The *rate* at which P is formed varies with time because the concentrations of A and B are reduced as the reaction progresses (Fig. 4.1). The form of the curve in Fig. 4.1 will depend on the number of reacting substances limiting the rate of reaction.

(A) *Zero-order reactions*

The reaction rate is independent of the concentration of reacting substances. (More often occurs in heterogeneous systems when factors such as surface area available for adsorption limit the reaction rate.)

If in the reaction $A + B = P$ the exponents v and w in equation (4.5) are zero, then the rate of change of concentration for each reactant may be represented by equation (4.6).

$$\frac{-d[A]}{dt} = \frac{-d[B]}{dt} = \frac{d[P]}{dt} = k \tag{4.6}$$

(B) *First-order reactions*

The reaction rate is proportional to the concentration of one of the reactants.

For either

$$A \to B + C$$

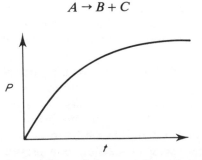

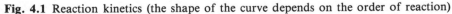

Fig. 4.1 Reaction kinetics (the shape of the curve depends on the order of reaction)

or

$$A + B = P$$

(where B is present in excess) only $[A]$ matters and the rate of reaction can be represented as

$$\frac{-\mathrm{d}[A]}{\mathrm{d}t} = k[A] \tag{4.7}$$

which can be integrated to give

$$[A] = [A_0]\, e^{-kt}$$
$$[A_0] = \text{initial concentration at } t = 0 \tag{4.8}$$

$[A]$ is the instantaneous concentration at time t and the amount of $[A]$ consumed in that period is

$$[A_0] - [A] = [A_0](1 - e^{-kt}) \tag{4.9}$$

By substituting $k_1 = 0.4343k$ equation (4.8) may be written in the form

$$[A] = [A_0]\, 10^{-k_1 t}$$

or

$$\log\frac{[A]}{[A_0]} = -k_1 t \tag{4.10}$$

k or k_1 can be evaluated from a plot of $\log [A]/[A_0]$ against t, which should give a straight line of slope k or k_1 (Fig. 4.2) using a least squares analyses. This is a useful method for determining whether a particular reaction is first order or not.

Values of k can also be determined from the time taken for half the initial

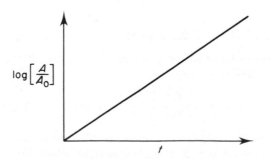

Fig. 4.2 Semi-logarithmic plot of a first-order reaction

material to react. Substituting the half-life concentration $[A_0]/2$ into equation (4.10) gives

$$\log_{10} 2 = -k_1 t_{1/2}$$

$$k_1 = 0.3/t_{1/2}$$

(4.11)

Examples of first-order reactions are (1) the decay of radio-isotopes; (2) the pseudo-first-order reactions of (a) disinfection of organisms in which the number of organisms destroyed per unit of time is proportional to the number of organisms remaining and (b) the hydrolysis of cane sugar in water solution according to

$$C_{12}H_{22}O_{11} + H_2O \rightarrow 2\ C_6H_{12}O_6$$

(4.12)

where water is present in such large excess that its concentration does not change significantly during the course of the reaction. The oxidation of organic matter and many other important reactions come into this category.

(C) Second-order reactions

Two types of second-order reactions can be identified

(a) $$A + B \rightarrow P$$

where the concentrations of both A and B affect the reaction rate which can be represented as

$$\frac{-d[A]}{dt} = k[A][B] = \frac{d[P]}{dt}$$

(4.13)

with values of both v and $w = 1$ in equation (4.5)

(b) $$A + A = P$$

giving

$$\frac{-d[A]}{dt} = k\ [A]^2 = \frac{dP}{dt}$$

(4.14)

$v = 2$ and $w = 0$ in equation (4.5).

For the second-order reaction in which both A and B react the integrated rate expression takes the form:

$$\log_{10} \frac{[B_0]\ ([A_0] - [A])}{[A_0]\ ([B_0] - [B])} = k_1([A_0] - [B_0])t$$

(4.15)

$[A]$ and $[B]$ the instantaneous concentrations are equal in a second-order reaction of this type. A plot of the LHS of equation (4.15) against t will yield a straight line with a slope equal to $k_1\ ([A_0] - [B_0])$.

Most reactions of significance in the environment can be approximated by the reaction orders described above however, higher orders and fractional orders are possible. Many reactions ordinarily observed are not simple processes but proceed in a series of steps, or in competition with other reactions, or in which the process is readily reversible. It is not possible to deduce the order of reaction of the rate limiting step from the stoichiometry of the equation. The order must be determined experimentally.

(D) Complex reactions

Three classes of complex reactions may be identified:
1. Consecutive reactions

$A \xrightarrow{k_1} B \xrightarrow{k_2} C$ two or more reactions occur in series

2. Back reactions

$$A + B \rightleftharpoons P \quad \text{readily reversible reaction}$$

3. Competing reactions

$$\left.\begin{array}{c} A + B \rightarrow P \\ A + C \rightarrow Q \end{array}\right\} \quad \text{occurring simultaneously}$$

Complete determination of the rate pattern for complex reactions involves simultaneous solution of the differential kinetic equations for each of the individual steps.

For consecutive reactions assuming first-order rates

$$\frac{-d[A]}{dt} = k_1 [A] \tag{4.16}$$

$$\frac{-d[B]}{dt} = k_2 [B] - k_1 [A] \tag{4.17}$$

$$\frac{d[C]}{dt} = k_2 [B] \tag{4.18}$$

Equations (4.16) to (4.18) may be integrated to give

$$[A] = [A_0] e^{-kt} \tag{4.19}$$

$$[B] = [A_0]\left(\frac{k}{k_2 - k_1} e^{-k_1^t} - e^{-k_2^t}\right) + [B_0] e^{-k_2^t} \tag{4.20}$$

$$[C] = [A_0]\left(1 - \frac{k_2 e^{-k_{1t}} - k_1 e^{-k_2^t}}{k_2 - k_1}\right) + [B_0] 1 - e^{-k_2^t} + [C_0] \tag{4.21}$$

$[A_0]$ $[B_0]$ and $[C_0]$ are the respective concentrations of $[A]$, $[B]$ and $[C]$ at time $t = 0$

If $k_1 \ll k_2$ the reactions will behave kinetically in a manner similar to a single-step reaction.

$$A \xrightarrow{k_1} C$$

When the process for a first-order reaction is readily reversible such as

$$A \underset{k_{-1}}{\overset{k_{+1}}{\rightleftharpoons}} B$$

the rate expression includes both the forward and reverse reactions

$$\frac{\mathrm{d}B}{\mathrm{d}t} = \frac{-\mathrm{d}A}{\mathrm{d}t} = k_{+1}[A] - k_{-1}[B] \tag{4.22}$$

at equilibrium

$$\frac{\mathrm{d}B}{\mathrm{d}t} = \frac{\mathrm{d}A}{\mathrm{d}t} = 0$$

and

$$\frac{[A]}{[B]} = \frac{k-1}{k+1} = K \qquad K = \text{equilibrium constant} \tag{4.23}$$

Comparable relationships can be determined for other complex reactions of different orders.

Temperature effect

Reaction rates generally increase with increase in temperature. Some will approximately double for each $10\,°C$ increase.

The Van't Hoff – Arrhenius equation can be used to predict the effect of temperature on the rate coefficient

$$\frac{\mathrm{d}(\log k)}{\mathrm{d}t} = \frac{E}{RT^2} \tag{4.24}$$

where

E = activation energy (calories) a constant characteristic of the reaction
R = gas constant (cal per $°C$)
T = absolute temperature.

Integrating (4.24) between the limits T_0 and T gives

$$\log \frac{k}{k_0} = \frac{E(T - T_0)}{RTT_0} \tag{4.25}$$

where k, k_0 are rate constants at T and T_0 respectively.

In aquatic systems with a small temperature change TT_0 does not change significantly and E/RTT_0 can be assumed constant. Equation (4.25) can be approximated by

$$k = k_0 \, e^{C_k(T - T_0)} \tag{4.26}$$

where C_k = temperature characteristic. Often the empirical form is used

$$k = k_0 \, \theta^{(T - T_0)} \tag{4.27}$$

where θ = temperature coefficient. Taking the first two terms of the expanded form of e^x equation (4.26) can be approximated to

$$k = k_0 \, (1 + \bar{C}_k \, (T - T_0)) \tag{4.28}$$

4.2.4 Heterogeneous systems

In heterogeneous systems, the rate of chemical reaction as described in the homogeneous system is further complicated by physical processes resulting from phase boundaries of incomplete mixing. At low concentrations, the system may be concentration dependent but at higher concentrations phase discontinuity may control the reactions.

Three phenomena can be identified: (a) osmosis, (b) diffusion, (c) adsorption.

(a) *Osmosis* can be defined as the passage of solvent through a membrane from a dilute solution into a more concentrated one. The rate of movement of molecules across the membrane is a function of the concentrations at each side. A net movement takes place until either the two concentrations are equal or until equilibrium is established by external pressure.

(b) *Diffusion* is the non-advective migration of a substance in solution or suspension in response to a concentration gradient of that substance through another substance. It is a basic natural process which accounts for most of the transport that takes place at the molecular level.

The mechanism is described by Fick's First Law which states that the rate of mass transport by diffusion across an element of unit area is proportional to the concentration gradient of the diffusing substance.

$$N_x = - D_m \frac{\partial c}{\partial x} \tag{4.29}$$

where

N_x = rate of mass transport across an element of area normal to x

$\dfrac{\partial c}{\partial x}$ = the concentration gradient of the diffusing phase in the x direction

Dm = the molecular diffusion coefficient which is proportional to the absolute temperature and inversely proportional to the molecular weight of the diffusing phase and the viscosity of the dispersing phase

Fick's Second Law describes the rate of change of concentration of the diffusing phase in an element of the dispersing phase.

$$\frac{\partial c}{\partial t} = Dm\, \frac{\partial^2 c}{\partial x^2} \tag{4.30}$$

or in three dimensions

$$\frac{\partial c}{\partial t} = Dm\left(\frac{\partial^2 c}{\partial x^2} + \frac{\partial^2 c}{\partial y^2} + \frac{\partial c}{\partial z^2} \right)$$

(c) *Adsorption* occurs when molecules in solution strike the surface of a solid absorbent and become attached to the surface. At low concentrations of absorbate the system may be concentration dependent. Under these conditions the rate of adsorption may be derived as follows:

$$\frac{dc}{dt} = k\phi(c) \tag{4.31}$$

where

c is the concentration of substance in solution
t is the time of exposure
$\phi(c)$ is a function of the concentration in solution
k is the rate coefficient

Assuming that $\phi(c)$ can be expressed as $C_0 - C$, the concentration adsorbed and that k may be modified to allow for the drop in adsorption response during the course of exposure giving

$$k = \frac{k_0}{1 + r^t}$$

k_0 = rate at $t = 0$
r = coefficient of retardation

then on integrating equation (4.31) and substituting for $\phi(c)$ and k

$$\frac{C}{C_0} = \left[1 - (1 + rt)^{-k_0/r} \right] \tag{4.32}$$

At higher concentrations of substance in solution the rate of deposition of molecules on a uniform solid surface may best be described by the Langmuir Isotherm which assumes the formation of a monomolecular layer.

Let x be the fraction of the total solid surface occupied by molecules, $(1 - X)$ being left free at any moment in time. The rate at which molecules are adsorped is proportional to the available surface area

$$\frac{dN_a}{dt} = k_1 C (1 - X)$$

and the rate of desorption is proportional to the surface covered

$$\frac{dN_d}{dt} = k_2 X$$

where k_1 and k_2 are rates of adsorption and desorption and N_a and N_d are the numbers of molecules adsorbed and desorbed. The dynamic equilibrium between free and adsorbed molecules can be expressed by

$$k_1 C (1 - X) = k_2 X \qquad (4.33)$$

Rearranging equation (4.33) gives

$$X = \frac{k_1 C}{k_2 + k_1 C} \qquad (4.34)$$

If the adsorbed molecules undergo a chemical transformation at a rate proportional to this surface density then

$$\text{the rate of transformation} = kX = \frac{k \cdot b \cdot C}{1 + bC}$$

where

$$b = \frac{k_1}{k_2}$$

For small values of C the *amount* adsorbed is linearly proportional to C while for large values of C the amount adsorbed is independent of C.

4.2.5 Enzyme reactions

The chemical reactions by which substances are synthesized into cellular materials are good examples of complex reactions taking place in a heterogeneous environment. Transformation of organic molecules in the synthesis of cellular materials will depend on the rate of diffusion of material to the cell wall, the rate of adsorption at the cell wall and the various rates of reaction at each step of the complex change within the cell wall. Intermediate compounds are formed which have no cellular function other than that of opening a pathway to the final cellular molecule and in a similar fashion many enzyme reactions take place before a complex organic substance is simplified and returned to more stable organic levels.

The resolution of these complex reactions into simple stages helps to determine the overall rate limiting step which effectively controls the cellular growth. Three advantages of resolution into stages are worth emphasizing.

(1) Simple individual steps may be independently influenced by conditions such as temperature and concentration, leading to a considerable diversity of products in various proportions giving an appearance of great complexity. Resolution into stages will reduce the apparent natural complexity to the level inherent in the system.

(2) The existence of transitory intermediates renders possible the coupling of reactions which appear to be independent from a stoichiometric viewpoint, e.g.

$$B + A_2 = BA_2 \qquad \text{with reduction in free energy}$$

$$X + Y = X + Y \qquad \text{with increase in free energy}$$

If the reduction of free energy in the former exceeds the increase in the latter then the reactions may be coupled as follows:

$$A_2 = A + A$$

$$A + XY = AX + Y$$

$$AX + B = AB + X$$

$$AB + A = BA_2$$

(3) Individual steps may be spatially separated and at a given moment in time not only concentrations of intermediates will exist but also concentration gradients will occur. The overall course of the change will depend on the space-time relationship in a variety of ways.

4.3 GROWTH KINETICS

The kinetics of growth are rather more complex than those of chemical reactions especially when dealing with organisms like invertebrates and fish. Fortunately, the growth of the most important group of micro-organisms—the bacteria—can be represented quite simply and those kinetics with suitable modification may also be used to describe the growth of algae.

For bacteria, the fundamental relationship is between the growth rate and the concentration of substrate. This is shown in Fig. 4.3 and may be expressed

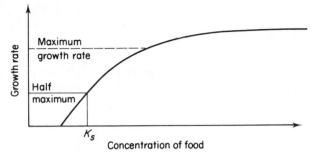

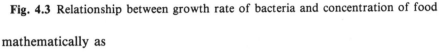

Fig. 4.3 Relationship between growth rate of bacteria and concentration of food

mathematically as

$$\mu = \mu_{\max} \left(\frac{S}{S + K_s} \right) \qquad (4.35)$$

where

μ = growth rate

S = substrate concentration

K_s = half-rate concentration as defined in Fig. 4.3

The other assumption is that there is a simple relationship between growth and utilization of the substrate which is defined by the yield.

Yield,

$$Y = \frac{\text{weight of bacteria formed}}{\text{weight of substrate utilized}}$$

$$\frac{\mathrm{d}X}{\mathrm{d}t} = - Y\frac{\mathrm{d}s}{\mathrm{d}t}$$

These two simplifying assumptions are not completely valid and later in Chapter 4 and Chapter 11 their application and modification are discussed further.

For algae, the same relationship between growth rate and nutrient concentration has been used but the connection is with an inorganic nutrient such as nitrogen, cphosphorous or silicon. The more fundamental process controlling growth is the balance between photosynthesis and respiration. The latter is easily described as a function of biomass:

$$\frac{\mathrm{d}R}{\mathrm{d}t} = R * A \qquad (4.37)$$

where

R = rate of respiration
A = algal concentration

but the rate of photosynthesis is also dependent on light intensity which varies with time and depth. The relationship with light is generally expressed in terms of the maximum rate of photosynthesis, P_{max}:

$$\frac{dP}{dt} = P_{max} \frac{I}{I_{opt}} \exp\left(1 - \frac{I}{I_{opt}}\right) \tag{4.38}$$

where

I = light intensity

I_{opt} = light intensity corresponding to P_{max}

Light intensity decreases with depth in an exponential manner at a rate which depends upon the turbidity, thus giving

$$I(z + h) = I(z) \exp(-Kh)$$

where

$I(z)$ and $I(z + h)$ = light intensities at depths z and $z + h$ respectively.

The death rate of algae is often represented by the rate of sedimentation since when algae die, they cease to be buoyant. Sedimentation is also a problem for some living algae e.g. the diatom Melosira which tends to sediment unless returned to the surface by advection.

Sedimentation may be represented as some function of the standing crop

$$\text{Rate of sed} = f(\text{algal concentration})$$

where f may have values around $1 - 10\%$ per day.

Other losses to the algal population are due to predation. Where several prey species are consumed by the same predator which eats indiscriminantly but only a fraction of the prey species A_i that it encounters then

$$\frac{dA_i}{dt} = \mu_i g A_i B \tag{4.39}$$

where

μ_i = fraction of species i

A_i = biomass of prey species i

B = biomass of predator

The growth kinetics of other groups of organisms such as invertebrates and fish may also need to be represented. This is usually done by dividing the population into a number of cohorts (or age groups) on the basis of their fecundity and chances of survival. A matrix of these values may then be used to multiply a vector of numbers in each cohort to determine the future population size and age structure:

$$
\begin{array}{ccc}
\text{Matrix of fecundity} & \text{Current} & \text{New population} \\
\text{and survival} & \text{population} = & \text{vector} \\
& \text{vector}
\end{array}
$$

$$
\begin{bmatrix} f_1 & f_2 & f_3 \\ p_1 & 0 & 0 \\ 0 & p_2 & 0 \end{bmatrix} \quad \begin{bmatrix} v_1 \\ v_2 \\ v_3 \end{bmatrix} \quad = \quad \begin{bmatrix} v_1 \\ v_2 \\ v_3 \end{bmatrix} \quad (4.40)
$$

$$
A \qquad\qquad V_1 \quad = \quad V_2
$$

The values of the fecundities f_1, f_2 and f_3 and the probabilities p_1 and p_2 are density dependent so that the population needs to be incorporated into their calculation e.g.

$$
A = \begin{bmatrix} 0 & 0.8F & 0.5F \\ 0.35 & 0 & 0 \\ 0 & 0.4S & 0 \end{bmatrix}
$$

where

$$
F = \frac{a}{b + N_t} \quad \text{and} \quad S = \frac{1}{1 + e^N t/c}
$$

where

$$
N_t = \text{Population density at time } t
$$

a, b and c are constants.

4.4 REACTOR THEORY

In the treatment of wastewaters there is a series of unit processes such as sedimentation, aeration, etc. which take place inside tanks or similar structures which may be regarded as chemical or biochemical reactors. Similarly, sections of an estuary, a lake or stretch of a river may be treated conceptually in the same manner. Reactor theory can therefore be applied widely in modelling water pollution and its control.

4.4.1 Types of reactor

The relationship between influent and effluent quality is determined by the dispersion characteristics of the reactor and the kinetics of any reactions. The two basic types of dispersion characteristics are as follows:

(a) Plug-flow where the liquid flows through the reactor without any mixing taking place;
(b) Completely mixed where liquid entering the reactor is instantaneously mixed with the entire contents.

These may be illustrated by the behaviour of a conservation tracer injected as a pulse into the influent of a reactor. The concentrations of tracer in the effluent have two typical shapes as shown in Fig. 4.4.

In practice, reactors are encountered that are in varying degrees intermediate between the two kinds. The dispersion characteristics of these intermediate types are most easily represented as a series of completely mixed reactors (e.g. see Chapter 12). The number of reactors in series can be obtained from the following equation:

$$E = \frac{C}{C_0} = \frac{j^j \, \theta^{(j-1)}}{(j-1)!} \, \exp(-j\theta) \qquad (4.41)$$

with mean

$$\theta_c = 1$$

and variance

$$\sigma_\theta^2 = \frac{1}{j}$$

$$\bar{t} = \frac{\Sigma t_c}{\Sigma C}$$

$$\sigma_t^2 = \frac{\Sigma t_c^2}{\Sigma C} - \frac{\Sigma t_c^2}{\Sigma C}$$

$$\sigma_\theta^2 = \frac{\sigma_t^2}{(\bar{t})^2}$$

where

C = effluent concentration

C_0 = initial effluent concentration

θ = $\dfrac{t}{t}$

j = number of equally-sized tanks

t = time

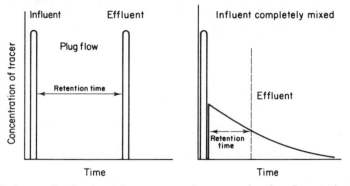

Fig. 4.4 Exit-age distribution of a conservative tracer in plug flow and completely mixed reactors

Alternatively, departure from ideality may be expressed by means of the dispersion number. This is based upon the idea of representing a reactor with some degree of dispersion by an equivalent ideal plug flow reactor with a correction factor i.e.

$$V_{\text{actual}} = V_{\text{plug flow}} \quad (\text{correction factor})$$

where the correction factor is some function of:

(i) Intensity of mixing;
(ii) Reactor geometry;
(iii) Reaction rate.

The intensity of mixing is generally expressed by the dimensionless group D/uL, where

D = dispersion coefficient
u = velocity
L = length of the reactor

This is the reciprocal of the Peclet number.

4.4.2 Plug-flow reactors

Plug-flow reactors may be visualized as containing a series of isolated volumes of liquid which are contiguous with their adjacent volumes but do not mix with them, rather like a series of carriages in a train.

The change of concentration during the retention time in the reactor is therefore dependent solely on processes within the individual plugs, and can therefore be represented by a batch reactor. For example, the bacterial and substrate kinetics in a batch reactor are shown in Fig. 4.6.

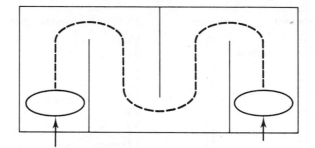

Fig. 4.5 Movement of material through a plug-flow reactor

Mathematically, these changes may be given by equation (4.42).

$$\frac{dX}{dt} = \mu_{max} \left(\frac{S}{S + K_s} \right) X - KdX \qquad (4.42)$$

where

  X = bacterial concentration
  μ = growth rate
  S = concentration of substrate
  Kd = death rate of bacteria

and the substrate kinetics by equation (4.43)

$$\frac{ds}{dt} = \frac{1}{Y} \mu_{max} \frac{S}{S + K_s} X \qquad (4.43)$$

where

$$Y = \text{yield}$$

Plug-flow reactors which have only chemical changes may also be described very simply by the appropriate zero order, first-order or second-order equation.

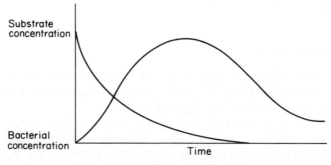

Fig. 4.6 Batch system kinetics

4.4.3 Completely mixed reactors

Completely-mixed reactors are slightly more complicated because bacteria and substrate are continuously being added and removed. They are best described by a mass balance approach. This can be illustrated by considering a completely mixed reactor in which a constituent is undergoing first order decomposition.

The mass balance model can be written as

$$\frac{dC_2}{dt} = \frac{Q_1 C_1}{V} - \frac{Q_2 C_2}{V} - KC_2 \qquad (4.44)$$

Rate of change of concentration = Flux in − Flux out − Rate of decomposition

When bacteria are involved two balances are required, one dealing with the organisms and the other describing the substrate.

The net rate of increase of concentration of organisms is given by the simple balance

$$\frac{dX}{dt} = \mu X - \frac{Q}{V} X - KdX \qquad (4.45)$$

Hence

$$\frac{dX}{dt} > 0 \quad \text{for } \mu > \left(\frac{Q}{V} + Kd \right) \quad \text{i.e. increase}$$

$$\frac{dX}{dt} = 0 \quad \text{for } \mu = \left(\frac{Q}{V} + Kd \right) \quad \text{i.e. steady state}$$

$$\frac{dX}{dt} < 0 \quad \text{for } \mu < \left(\frac{Q}{V} + Kd \right) \quad \text{i.e decrease}$$

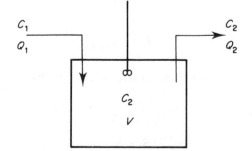

Fig. 4.7 Mass balance model of completely mixed reactor with reaction

The change in concentration of organisms is related to the substrate balance by

$$\mu = \mu_{max} \left(\frac{S}{S + K_s} \right) \qquad (4.46)$$

and so the substrate balance is coupled to the organism balance. The substrate balance may be expressed by

$$\frac{dS_2}{dt} = \frac{Q}{V} * S_1 - \frac{Q}{V} * S_2 - \frac{\mu X}{Y} \qquad (4.47)$$

Substituting for μ the two mass balances may be written as follows:

$$\frac{dX}{dt} = X \left[\mu_m \left(\frac{S_2}{K_s + S_2} \right) - \frac{Q}{V} - Kd \right] \qquad (4.48)$$

$$\frac{dS}{dt} = \frac{Q}{V} (S_1 - S_2) - \frac{\mu_m X}{Y} \left(\frac{S_2}{K_s + S_2} \right) \qquad (4.49)$$

It is apparent from these equations that if S_1, Q and V are constant and that Q/V does not exceed a critical value then a steady-state will be achieved at which

$$\frac{dX}{dt} = \frac{dS}{dt} = 0$$

The tendency for this steady-state to be attained may be explained as follows:

Initially X is very small
therefore

$$S_2 \simeq S_1$$

and

$$\mu > \left(\frac{Q}{V} + Kd \right)$$

The concentration of organisms will therefore increase causing a decrease in S_2 so that eventually

$$\mu = \left(\frac{Q}{V} + Kd \right)$$

The steady-state solution to these equations is given by

$$\bar{S} = \frac{K_s(1 + \theta Kd)}{\theta(\mu_{max} - Kd) - 1} \qquad (4.50)$$

$$\bar{X} = \frac{Y(S_1 - S_2)}{1 + \theta Kd} \qquad (4.51)$$

where

$$\theta = \frac{Q}{V}$$

The effect of varying conditions in the reactor effluent can be considered in terms of flow rate and influent concentration. Increased flow rate reduces the retention time and initially washes more organisms out of the reactor and initially increases the substrate level. The growth rate increases due to the increase in substrate and restores the bacterial concentration to its previous concentration. These changes are illustrated graphically in Fig. 4.8. If plotted on appropriate scales the curves for bacteria and substrate are mirror images.

Figure 4.8 also shows the limit on possible steady-states as the dilution rate $(D = Q/V)$ plus the death rate (Kd) approach the maximum growth rate. The critical values at which washout occurs is obviously of great concern. The value of the dilution rate D_c is equal to the highest possible value of the growth rate μ which is attained when $S_1 = S_2$. It is given by equation (4.52)

$$Dc = \mu_m \frac{S_1}{K_s + S_1} - Kd \tag{4.52}$$

When

$$S_1 \gg K_s \quad \text{which is usual then } D_c = \mu_{\max}.$$

The effect of different substrate concentrations in the influent to a reactor operated at varying dilution rates is shown in Fig. 4.9. It can be seen that at a given dilution rate, below D_c, the concentration of bacteria is nearly proportional to S_1 but that the effluent substrate concentration is independent of S_1, i.e. for a given value of D the concentration of S_2 comes to an equilibrium such that $\mu = D + Kd$ and this is independent of S_1. The curve for X against D shifts vertically for changes in S_1. Also the slope of the X curve near D_c is less steep for lower values of S_1. Consideration of the above equations shows that the

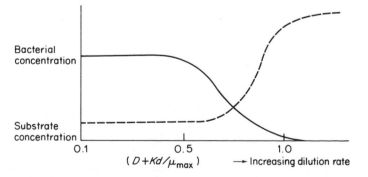
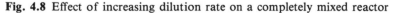

Fig. 4.8 Effect of increasing dilution rate on a completely mixed reactor

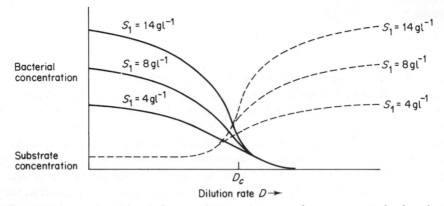

Fig. 4.9 Effect of varying influent substrate concentration on a completely mixed reactor

important factor is not the absolute value of S_1 but the ratio S_1/K_s. The higher this ratio, the greater the fraction of total substrate that can be consumed without an appreciable decrease in specific growth rate. Hence, as S_1/K_s is increased, the concentration of organisms is maintained at nearly the maximum possible value of D.In modelling the kinetics of completely mixed reactors it is important to be able to determine the values of the maximum growth rate, death rate, yield and Michaelis-Menten coefficient. These may be obtained graphically from experimental data by the following procedure:

(a) Plot the function $(S_1 - S_2)/X$ against the reciprocal of dilution rate. This should give a straight line, *the* slope of which is Kd/Y and the vertical intercept is $1/Y$ as shown in Fig. 4.10.

(b) Plot the function $\theta/(1 + \theta Kd)$ against the reciprocal of S_2. This should give a straight line whose slope is Ks/μ_{max} and whose intercept is $1/\mu_{max}$.

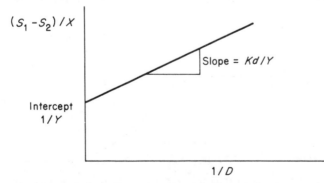

Fig. 4.10 Graphical method for estimating yield and death rate from continuous culture experiments

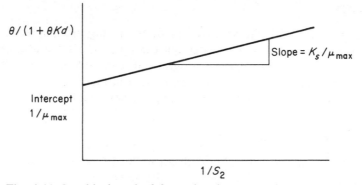

Fig. 4.11 Graphical method for estimating K_s maximum growth rate

At steady-state

$$D = \mu_{max} \left(\frac{S_2}{K_s + S_2} \right) = \frac{1}{\theta} \qquad (4.53)$$

which can be rewritten as

$$\frac{1 + \theta Kd}{\theta} = \mu_{max} \left(\frac{S_2}{K_s + S_2} \right) \qquad (4.54)$$

which may be rearranged to give

$$\frac{\theta}{1 + \theta Kd} = \left(\frac{K_s}{\mu_{max}} \right) \left(\frac{1}{S_2} \right) + \frac{1}{\mu_{max}} \qquad (4.55)$$

As shown in Fig. 4.11, a plot of $\theta/(1 + \theta Kd)$ against the reciprocal of S_2 gives a straight line, the slope of which is K_s/μ_{max} and the vertical intercept is equal to $1/\mu_{max}$.

4.5 BIBLIOGRAPHY

Sawyer, C.N. and McCarty, P.L. (1978). *Chemistry for Environmental Engineers*. McGraw-Hill, New York.

Levenspiel, O. (1962). *Chemical Reaction Engineering*. John Wiley & Sons, New York.

Atkinson, B. (1974). *Biochemical Reactors*. Pion, London.

Rich, L.G. (1973). *Environmental Systems Engineering*. McGraw-Hill, New York.

Morley, D.A. (1979). *Mathematical modelling in water and wastewater treatment*. Applied Science Publishers, Barking, Essex.

APPENDIX

Before proceeding with the applications of modelling in Chapters 5 – 14, it may be helpful to look at some very simple examples of batch and continuous-flow reactors. For each problem a program of a suitable model is listed but this is not intended to be the sole or definitive solution. Modelling is as much an art as a science and many different approaches could be used.

Example 1—Batch reactor

A batch reactor begins operation with an initial BOD of 500 mg 1^{-1} and a bacterial concentration of 200 mg 1^{-1}. If the maximum growth rate is 0.4 per hour, the death rate 0.1 per hour, the yield is 0.4 and the value of K_s is 300 mg 1^{-1}, write a program to print out the concentration of BOD and bacteria with time until the BOD falls below 20 mg 1^{-1}.

```
10   REM BATCH MODEL
20   XØ = 200
30   SØ = 500
40    U = 0.4
50   K1 = 300
60   K2 = 0.1
70    Y = 0.4
80   FOR T = 1 TO 1000
90   X1 = U * (SØ/(K1 + SØ)) * XØ – K2 * XØ
100  S1 = (U/Y) * (SØ/(K1 + SØ)) * XØ
110  XØ = XØ + X1
120  SØ = SØ + S1
130  IF SØ <  20 THEN 200
140  PRINT "BOD = ", SØ
150  PRINT "BACTERIA = ", XØ
160  NEXT T
170  END
```

Example CSTR

Write a program to calculate the steady-state concentrations of effluent BOD and bacteria from a stirred-tank reactor of capacity 400 m^3 using the same data as in Example 1. Assume that the inflow varies between 20 m^3 per hour for odd hours and 50 m^3 per hour for even hours and that the outflow is constant at 35 m^3 per hour.

```
10   CSTR MODEL
20   XØ = 200
30   SØ = 500
```

```
 40   U = 0.4
 50   K1 = 300
 60   K2 = 0.1
 70   Y = 0.4
 80   V = 400
 90   S1 = SØ
100   X1 = XØ
110   FOR N = 1 TO 100
120   FOR T = 1 TO 2
130   READ Q1, Q2
140   V1 = Q1 − Q2
150   S2 = Q1/V * SØ − Q1/V * S1 − (U/Y) * (S1/(S1 + K1))
155   IF S2 < 0.1 THEN 260
160   X2 = Q1/V * XØ − Q1/V * X1 − U * (S1/(S1 + K1)) * X1
165   IF X2 < 0.1 THEN 260
170   V = V + V1
180   X1 = X1 + X2
190   S1 = S1 + S2
200   PRINT "BOD = ", S1
210   PRINT "BACTERIA = ", X1
220   NEXT T
230   RESTORE
240   NEXT N
250   DATA 20, 35, 50, 35
260   END
```

Chapter 5

Models of Water Quality in Rivers

D. J. ELLIOTT AND A. JAMES

5.1 INTRODUCTION

Rivers have traditionally been used for the disposal of domestic and industrial wastewaters. In many cases, this has caused undesirable changes to the aquatic flora and fauna. The majority of these changes have been brought about by the discharge of organic matter (BOD) resulting in the lowering in the concentration of the dissolved oxygen (DO) in the receiving water. Pollution of rivers and estuaries is also frequently caused by the discharge of toxic substances, which may break down due to chemical or bacterial action (non-conservative) or which may be resistent to breakdown (conservative) and other problems may arise due to the discharge of inorganic nutrients causing excessive algal growth.

In all of these situations it is important to be able to relate the rate of discharge of the pollutant to resulting concentration pattern in the receiving water. Various methods have been devised for calculating the pattern beginning with the classic work on BOD/DO models by Streeter and Phelps in the 1920s. This laid the basis for modelling the chemical kinetics of breakdown. Subsequent work has concentrated on the hydrodynamic aspects—advection and diffusion along with work on stochastic and statistical models, and refinement of the kinetic models.

The following notes discuss the hydrodynamic basis for models of rivers followed by examples of their application.

5.2 THE CONVECTIVE DIFFUSION EQUATION

The convective diffusion equation has been used as the basis for many river and estuary models and is often used as the starting point in the literature in descriptions of model development. Before considering the equation in detail it is useful to briefly look at its theoretical development.

The total mass of material entering into an element of space in a given time must equal the increase in mass within the space in that time.

Figure 5.1 represents an element of space fixed in space relative to the Earth

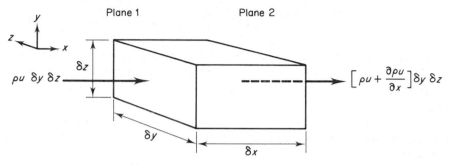

Fig. 5.1 Conservation of mass

within a fluid stream. The principle of conservation of matter applies to the fluid itself and also to material dissolved in the fluid or suspended in the fluid. Assume initially that a fluid with density ϱ flows through the box and that the fluid velocity has components u, v and w in the x, y and z directions.

The flux of material passing Plane 1 is equal to density x velocity x area $= \varrho u \, \delta y \, \delta z$.

Because the elemental volume is considered to be infinitesimally small it is possible to use Taylor's expansion to describe the flux passing Plane 2, i.e.

$$\left[\varrho u + \frac{\partial \varrho u}{\partial x} \, \delta x \right] \delta y \, \delta z$$

Hence the net decrease in mass due to flow in the x direction in the time δt

$$\frac{\partial (\varrho u)}{\partial x} \, \delta x \, \delta y \, \delta z \, \delta t$$

Similar expressions for the y and z directions may be obtained.

Assuming the initial mass in the element at time t is $\varrho \, \delta x \, \delta y \, \delta z$, then the mass at time $t + \delta t$ can again be obtained from a Taylor expression as

$$\left[\varrho \frac{\partial \varrho}{\partial t} \, \delta t \right] \delta x \, \delta y \, \delta z$$

Giving the rate of change of mass within the element as

$$\frac{\partial \varrho}{\partial t} \, \delta t \, \delta x \, \delta y \, \delta z$$

Equating the three mass flux terms with the time rate of change gives

$$-\frac{\partial \varrho}{\partial t} \, \delta t \, \delta x \, \delta y \, \delta z = \left[\frac{\partial (\varrho u)}{\partial x} + \frac{\partial \varrho v}{\partial y} + \frac{\partial \varrho w}{\partial z} \right] \delta x \, \delta y \, \delta z \, \delta t \qquad (5.1)$$

$$-\frac{\partial \varrho}{\partial t} = \frac{\partial (\varrho u)}{\partial x} + \frac{\partial (\varrho v)}{\partial y} + \frac{\partial (\varrho w)}{\partial z}$$

$$-\frac{\partial \varrho}{\partial t} = \varrho \left(\frac{\partial u}{\partial x} + \frac{\partial v}{\partial y} + \frac{\partial w}{\partial z}\right) + \frac{u\partial \varrho}{\partial x} + \frac{v\partial \varrho}{\partial y} + \frac{w\partial \varrho}{\partial x} \quad (5.2)$$

In equation (5.2) $\partial/\partial t$ is often called the local term which describes changes that would occur independent of the movement of the fluid particle. In effect, it is the rate of change that would occur for a motionless particle at a certain point.

The terms $u(\partial/\partial x) + v(\partial/\partial y) + w(\partial/\partial z)$ are 'convective' terms which describe the rate of change due to the particle moving in a field where gradients of the property exist.

The combined effect of $\partial/\partial t + u(\partial/\partial x) + v(\partial/\partial y) + w(\partial/\partial z)$ is often referred to as the substantial or total derivative in fluid mechanics and represents the total rate of change of some property of the fluid experienced by a particular particle of fluid as it moves with velocity components u, v and w.

The total derivative may be expressed as

$$\frac{D}{Dt} = \frac{\partial}{\partial t} + u\frac{\partial}{\partial x} + v\frac{\partial}{\partial y} + w\frac{\partial}{\partial z} \quad (5.3)$$

An example may help to clarify the physical significance of these terms.

Assume an aeroplane is flying south (x direction) at a constant height. Throughout the area of flight the temperature is increasing by $1\,^{\circ}$C per day. Also for each 1000 miles of travel the temperature increased by $2\,^{\circ}$C.

Thus the local effect experienced by a stationary body would be an increase in temperature of $1\,^{\circ}$C/day.

The total rate of increase experienced by the aircraft is

$$\frac{DT}{Dt} = \frac{\partial T}{\partial t} + u\frac{\partial T}{\partial x} = 1 + \frac{u \times 2}{1000}$$

where T is the mean daily temperature. If the velocity u is 200 miles per hour, then the total rate of increase experienced by the plane is

$$\frac{DT}{Dt} = 1 + 200 \times 24 \times \frac{2}{1000}$$

$$= 1 + 9.6 = 10.6\,^{\circ}\text{C/day} \quad (5.4)$$

Returning to equation (5.2), this may now be expressed as

$$\frac{D\varrho}{Dt} + \varrho\left(\frac{\partial u}{\partial u} + \frac{\partial v}{\partial y} + \frac{\partial w}{\partial z}\right) = 0$$

This equation is known as the continuity equation which may be further modified for an incompressible fluid for which $D\varrho/Dt = 0$. That is the density of the fluid remains constant over time which is generally true for normal river and estuary problems. Equation (5.5) then becomes

$$\frac{\partial u}{\partial x} + \frac{\partial v}{\partial y} + \frac{\partial w}{\partial x} = 0 \qquad (5.6)$$

which is the continuity equation for an incompressible fluid in steady or unsteady flow.

Although changes in density may be considered insignificant over the depths encountered in rivers and estuaries the concentrations of salts in solution or solids in suspension may change rapidly over the distances involved.

Movement with the bulk water flow, advection, is not the only physical mechanism which must be considered and random movements due to turbulence of the fluid or interchange of material between fast- and slow-moving layers of fluid must also be considered. These processes cause dispersion of material suspended in the liquid mass. This dispersion is relative to the bulk liquid flow and may be described mathematically in a form analagous to the molecular diffusion process.

5.3 MOLECULAR DIFFUSION AND FICK'S LAW

Assume that a conservative material is suspended or dissolved in a liquid.

Fick's First Law states that the rate of mass transport of material or flux through the liquid, by molecular diffusion is proportional to the concentration gradient of the material in the liquid.

$$\text{Diffusive mass flux} = -\,\text{Dm}\,\frac{\partial c}{\partial x}$$

where Dm is the molecular diffusion coefficient or constant of proportionality. It is proportional to absolute temperature and also inversely proportional to the molecular weight of the diffusing phase and the viscosity of the dispersing phase. The negative sign is an indication that material flows from areas of high concentration to areas of low concentration due to the diffusive process.

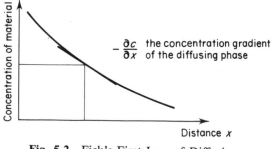

Fig. 5.2 Fick's First Law of Diffusion

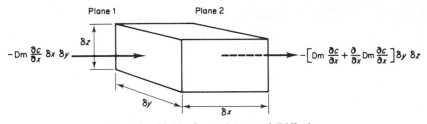

Fig. 5.3   Fick's Second Law of Diffusion

Ficks's Second Law may be obtained by again considering an elemental volume fixed in space in which the transport mechanism is pure diffusion. By equating the time rate of change of mass within the element to the change caused by the diffusive flux in each of the three coordinate directions

$$\frac{\partial c}{\partial t} = -\frac{\partial}{\partial x}\, Dm\, \frac{\partial c}{\partial x} - \frac{\partial}{\partial y}\, Dm\, \frac{\partial c}{\partial y} - \frac{\partial}{\partial z}\, Dm\, \frac{\partial c}{\partial z}$$

and assuming that the molecular diffusion coefficient Dm is constant in all coordinate directions

$$\frac{\partial c}{\partial t} = -Dm \left[\frac{\partial^2 c}{\partial x^2} + \frac{\partial^2 c}{\partial y^2} + \frac{\partial^2 c}{\partial z^2}\right] \tag{5.7}$$

Equation (5.7) describes the rate of change in concentration with respect to time, of a substance subject only to the molecular diffusion process and the use of the constant of proportionality Dm is restricted to this context.

The dispersion of effluents subject to large-scale processes of advection and turbulence may be described mathematically by an analogy with Fick's Laws of diffusion.

Consider Fig. 5.4 in which the elemental volume is fixed in a fluid medium. The conservation of mass equation of a substance introduced into the fluid can be obtained in a similar way to that used for the continuity equation.

Neglecting molecular diffusion relative to the large-scale mixing due to tur-

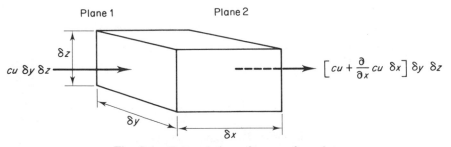

Fig. 5.4   Conservation of mass of a solute

bulence, the flux of material into the element across Plane 1 is $cu\,\delta y\,\delta z$ where u is the instantaneous velocity in the x direction.

The net change in mass of material in the element from the flux in the x direction is $-(\partial/\partial x)cu\,\delta x\,\delta z$.

Equating the time rate of change of change of mass in the element with the rate of change due to the flux in each of the three coordinate directions gives

$$\frac{\partial c}{\partial t} + \frac{\partial}{\partial x}(cu) + \frac{\partial}{\partial y}(cv) + \frac{\partial}{\partial z}(cw) = 0 \tag{5.8}$$

The values of concentration and velocity although in theory are instantaneous are in practice measured over short periods of time. Thus the instantaneous velocity component u can be expressed as

$$u = \bar{u} + u'$$

where

$\bar{u}$ is a time averaged component $= \dfrac{1}{T}\displaystyle\int_{t}^{t+T} u\,dt$ $\qquad$ (5.9)

and T is the period of measurement.

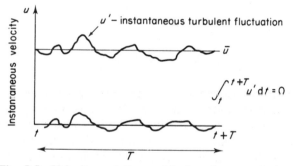

Fig. 5.5 Velocity variations with time due to turbulence

Also

$$v = \bar{v} + v'$$

$$w = \bar{w} + w'$$

$$c = \bar{c} + c'$$

These expressions for instantaneous velocities can be substituted into equation (5.8) and each term averaged to give

$$\frac{\partial}{\partial t}(\overline{\bar{c} + c'}) + \frac{\partial}{\partial x}(\overline{\bar{c} + c'})(\overline{\bar{u} + u'}) + \frac{\partial}{\partial y}(\overline{\bar{c} + c'})(\overline{\bar{v} + v'})$$

$$+ \frac{\partial}{\partial z}(\overline{\bar{c} + c'})(\overline{\bar{w} + w'}) = 0 \tag{5.10}$$

The terms in brackets may be expanded and all terms with only one prime will average to zero. For example, a term such as $\bar{c}u' = 0$.

Then

$$\frac{\partial \bar{c}}{\partial t} + \frac{\partial}{x}(\overline{\bar{u}\bar{c}}) + \frac{\partial}{\partial y}(\overline{\bar{v}\bar{c}}) + \frac{\partial}{\partial z}(\overline{\bar{w}\bar{c}}) + \frac{\partial}{\partial x}(\overline{u'c'})$$

$$+ \frac{\partial}{\partial y}(\overline{v'c'}) + \frac{\partial}{\partial z}(\overline{w'c'}) = 0 \qquad (5.11)$$

The differentials of products may be expanded in the form $u(dv/dx) + v(dn/dx)$ giving

$$\frac{\partial \bar{c}}{\partial t} + \bar{u}\frac{\partial \bar{c}}{\partial x} + \bar{v}\frac{\partial \bar{c}}{\partial y} + \bar{w}\frac{\partial \bar{c}}{\partial z} + \bar{c}\left(\frac{\partial \bar{u}}{\partial x} + \frac{\partial \bar{v}}{\partial y} + \frac{\partial \bar{w}}{\partial z}\right)$$

$$+ \frac{\partial}{\partial x}(\overline{\bar{u}c'}) + \frac{\partial}{\partial y}(\overline{v'c'}) + \frac{\partial}{\partial z}(\overline{w'c'}) = 0 \qquad (5.12)$$

Using the continuity equation

$$\frac{\partial \bar{u}}{\partial x} + \frac{\partial \bar{v}}{\partial y} + \frac{\partial \bar{w}}{\partial z} = 0$$

Equation (5.12) simplifies to

$$\frac{\partial \bar{c}}{\partial t} + \bar{u}\frac{\partial \bar{c}}{\partial x} + \bar{v}\frac{\partial \bar{c}}{\partial y} + \bar{w}\frac{\partial \bar{c}}{\partial z} + \frac{\partial}{\partial x}(\overline{u'c'}) + \frac{\partial}{\partial y}(\overline{v'c'}) + \frac{\partial}{\partial z}(\overline{w'c'}) = 0 \quad (5.13)$$

The cross-product terms such as $u'c'$ represent the net convection of mass due to the turbulent fluctuations and by analogy with Fick's Law of molecular diffusion they can be represented by an equivalent diffusive mass transport system in which the mass flux is proportional to the mean concentration gradient and the flux is in the direction of the mean concentration gradient. Hence

$$\overline{u'c'} = -Dx\frac{\partial \bar{c}}{\partial x}$$

$$\overline{v'c'} = -Dy\frac{\partial \bar{c}}{\partial y}$$

$$\overline{w'c'} = -Dw\frac{\partial \bar{c}}{\partial z}$$

where Dx, Dy and Dz are coefficients of turbulent diffusion. Equation (5.13) can now be rewritten omitting the time average bars as

$$\frac{\partial c}{\partial t} + u\frac{\partial c}{\partial x} + v\frac{\partial c}{\partial y} + w\frac{\partial c}{\partial z} - \frac{\partial}{\partial x}\left(Dx\frac{\partial c}{\partial x}\right) - \frac{\partial}{\partial y}\left(Dy\frac{\partial c}{\partial y}\right) - \frac{\partial}{\partial z}\left(Dz\frac{\partial c}{\partial z}\right) = 0$$

$$(5.14)$$

Dx, Dy and Dz are not necessarily the same in all directions. However, they will be several orders of magnitude greater than Dm the molecular diffusion coefficient. They may also change in value depending upon the position in the stream or estuary because of the change in stream dimensions.

Equation (5.14) is the three-dimensional convective diffusion equation which in this general form has no analytical solution and is extremely difficult to approximate numerically for use in a computer model. Values for the turbulent diffusion coefficient must be obtained from tracer measurements or by a curve fitting exercise and some way must be found for specifying the velocity components over time. This may be done by first solving the Momentum equation and using the resulting velocities in the convective diffusion equation or by calculating velocities from volume changes in the case of estuaries subject to tidal fluctuations. Whichever method is used, a knowledge of freshwater flow is required over the period of time being considered. For these reasons simplifying assumptions are usually made to reduce equation (5.14) to a form amenable for solution.

Equation (5.14) may be reduced to a one-dimensional form by taking average values of all parameters across the section of the stream or estuary and considering concentration only as a function of x and t.

For example,

$$\left. \begin{array}{l} U = \dfrac{1}{A} \displaystyle\int_0^A \bar{u} \; dA \\[4mm] C = \dfrac{1}{A} \displaystyle\int_0^A \bar{c} \; dA \end{array} \right\} \tag{5.15}$$

where A is the cross-sectional area.

$$\frac{\partial c}{\partial y} \quad \text{and} \quad \frac{\partial c}{\partial z} = 0, \qquad v = w = 0$$

and equation (5.14) reduces to

$$\frac{\partial c}{\partial t} + U \frac{\partial c}{\partial x} = \frac{1}{A} \frac{\partial}{\partial x} \left(EA \frac{\partial c}{\partial x} \right) \tag{5.16}$$

where E is now an effective longitudinal dispersion coefficient.

Equation (5.16) has been used for both rivers and estuaries and under certain assumptions may be simplified even further.

5.4 RIVER MODELS

Equation (5.16) was derived from first principles assuming an existing concentration of a conservative substance in the fluid. For most practical applications equation (5.16) must be extended to include source and sink terms for polluting discharges. For example, assume that the material being

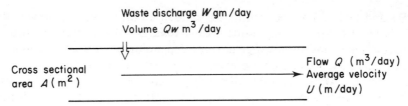

Waste discharge W gm/day

Volume Qw m^3/day

Cross sectional
area A(m^2)

Flow Q (m^3/day)
Average velocity
U (m/day)

Fig. 5.6 Conceptual diagram of a river model

discharged is subject to first-order reaction kinetics

$$\frac{dc}{dt} = \pm kc$$

where k is a first-order rate coefficient and the positive and negative signs indicate that the material may accumulate or disappear due to chemical or biochemical reactions.

Equation (5.16) becomes

$$\frac{\partial c}{\partial t} + U\frac{\partial c}{\partial x} - \frac{1}{A}\frac{\partial}{\partial x}\left(EA\frac{\partial c}{\partial x}\right) \pm kC + La = 0 \tag{5.17}$$

where La is the rate of addition of material in ppm/day at the point of discharge, assuming an initial dilution of the effluent load into the finite volume. La may be computed by a mass balance at the outfall

$$La = \frac{W}{Q + Qw}$$

Equation (5.17) is a second-order parabolic partial differential equation which may be solved analytically or by numerical methods.

5.4.1 Analytic solutions

Particular solutions of equation (5.17) must satisfy the given differential equation and also comply with given initial and boundary conditions.

The initial conditions specify the values of La and C as functions of x along the stretch of interest at time t = 0. The boundary conditions specify the values of L and C as functions of time at the beginning and end of the stretch. Boundaries must be chosen at sufficient distance from the stretch under consideration for the solution to closely approximate the specified conditions.

Continuous discharges

One form of analytic solution for equation (5.17) assumes that the discharge is continuous and the system has reached steady state, i.e. $\partial c/\partial t = 0$.

Also the stretch under consideration is assumed to have uniform flow and uniform cross-sectional area and the material being discharged does not significantly affect the flow in the river.

Equation (5.17) now becomes

$$E \frac{\partial^2 c}{\partial x^2} - U \frac{\partial c}{\partial x} \pm kc = 0 \qquad (5.18)$$

assuming a first-order decay term.

The steady-state solution for this situation is given by

$$C = \frac{W}{AUm} \exp\left(\frac{U}{2E} [1 \pm m] x\right) \qquad (5.19)$$

where

$$m = \sqrt{1 + 4\frac{kE}{U^2}}$$

In equation (5.19) the negative value of the exponent applies to the region down stream of the point of discharge. The positive exponent applies upstream.

Changes in the relative magnitudes of E and k in equation (5.19) will produce different answers as shown below.

In Fig. 5.7, $k = 0$ (i.e. a conservative substance) the concentration at the discharge point is equal to W/AU the rate of addition of the substance divided by the rate of flow in the river. The concentration is constant downstream of the outfall and decreases exponentially upstream.

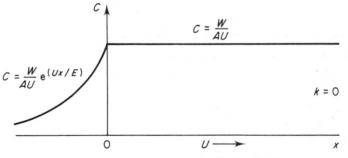

Fig. 5.7 Concentration profile for $k = 0$

In Fig. 5.8, $E = 0$, no transport upstream occurs. Downstream the concentration decays from an initial value of W/AU at the point of discharge governed by

$$C = \frac{W}{AU} e - \left(\frac{kx}{u}\right)$$

In Fig. 5.9a both E and k are non zero. The initial concentration is given by $C = W/AUm$ and decays both upstream and downstream. In practice, because of the unidirectional nature of E there is no significant transport upstream of the point of discharge and the initial concentration is essentially

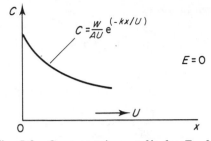

Fig. 5.8 Concentration profile for $E = 0$

equal to W/AU. This is because most dispersion is caused by velocity shear rather than turbulent eddies so that the upstream dispersion mechanism is much smaller than the downstream advection of the bulk water flow.

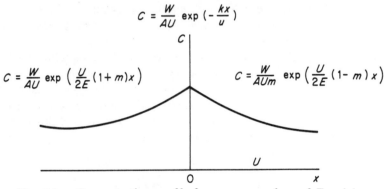

Fig. 5.9a Concentration profile for non-zero values of E and k

Another form of particular solution to equation (5.17) is the situation where W is an instantaneous conservative discharge again to a stream of uniform cross sectional area.

The release at $t = 0$ and $x = 0$ produces a Gaussian concentration distribution with respect to x. The centre of the distribution moves downstream at velocity U as shown in Fig. 5.9b.

The solution to this equation is

$$C = \frac{W}{A\sqrt{4\Pi\, Et}} \exp\left[-\frac{(x - Ut^2)}{4Et} \right] \tag{5.20}$$

Where

$\quad W$ = weight of conservative substances
$\quad A$ = cross-sectional area
$\quad t$ = time
$\quad x$ = distance downstream
$\quad U$ = mean velocity
$\quad E$ = dispersion coefficient

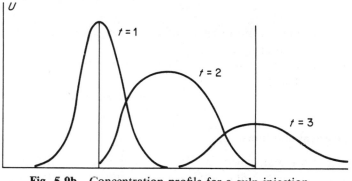

Fig. 5.9b Concentration profile for a gulp injection

When there is no advection i.e. if the discharge were into a canal the solution is

$$C = \frac{W}{A\sqrt{4\Pi\, Et}}\, \exp\left(\frac{-x^2}{4Et}\right) \qquad (5.20)$$

In this case the concentration profiles are those shown in Fig. 5.10.

It is possible to estimate the value of E from tracer studies in which a slug of tracer is injected into the river. The time concentration curve of the tracer is measured at two stations downstream of the injection point. E is obtained from

$$E = \frac{U^2}{2}\, \frac{(\sigma_1^2 - \sigma_2^2)}{(t_2 - t_1)}$$

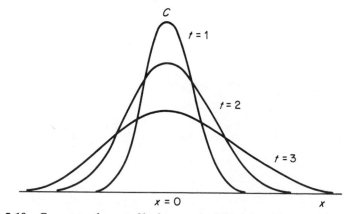

Fig. 5.10 Concentration profile for a gulp injection with zero advection

where

> t_1, t_2 are the mean times of the passage of the tracer past each station
>
> σ_1^2 and σ_2^2 are the variances of the time concentration curves at stations 1 and 2
>
> U is the mean velocity of flow between stations

5.4.2 Numerical solutions

It is often more convenient to solve the convective diffusion equation numerically using either finite difference or finite element techniques. Finite differences have been extensively used as the basis of river models with a variety of schemes being adopted. Essentially, the numerical approach attempts to approximate the continuous solution at a discrete number of points in time and distance. For example, the one dimensional equation (5.17) may be approximated on a time distance plane as shown in Fig. 5.11. At each point in time represented by the j rows the concentration of the parameter of interest must be evaluated at each distance mesh point represented by i columns.

An example of this form of solution is demonstrated using an explicit finite difference scheme. The solution must satisfy certain initial and boundary conditions. At time $t = 0$ concentration levels must be specified at each distance mesh point. Also, sufficient distance mesh points must be chosen so that the solution in the stretch of interest is not affected by the specified boundary conditions. Values at the boundaries are often chosen to represent natural river conditions.

In order to simplify the notation, equation (5.17) may be modified by

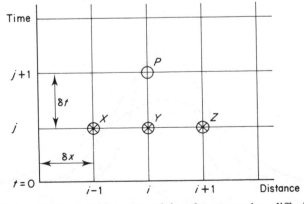

Fig. 5.11   Finite difference scheme for solving the convective–diffusion equation

assuming a constant cross-sectional area and a conservative pollutant to give

$$\frac{\partial c}{\partial t} = - U \frac{\partial c}{\partial x} + E \frac{\partial^2 c}{\partial x^2} + La \qquad (5.21)$$

In explicit difference form equation (5.21) may be written as

$$\frac{C_{i,j+1} - C_{i,j}}{\delta t} = - \frac{U}{\delta x} \left(C_{i,j} - C_{i-1,j} \right) + \frac{E}{\delta x^2} \left(C_{i-1,j} \right.$$

$$\left. + 2C_{i,j} + C_{i+1,j} \right) + La \qquad (5.22)$$

Where La is the concentration in ppm which must be added to each mesh point at which a discharge takes place. The value of La is obtained from a mass balance of the effluent load and the river flow at the point of entry. It is convenient for the effluent discharge points and the mesh points to coincide. If not it is necessary to interpolate between adjacent mesh points.

Inspection of equation (5.22) shows some interesting features. Firstly, let us assume that dispersion is negligible in relation to advection. Thus $E = 0$ and equation (5.22) reduces to a purely advective equation. Considering Fig. 5.12, it can be seen that for pure advection to be described correctly by the difference scheme the element of water at the point $i - 1$ at time j must move to i at time $j + 1$.

Thus

$$C_{i,j+1} = C_{i-1,j} \qquad (5.23)$$

For equation (5.22) to satisfy this requirement it is necessary that

$$U \, \delta t = \delta x \qquad (5.24)$$

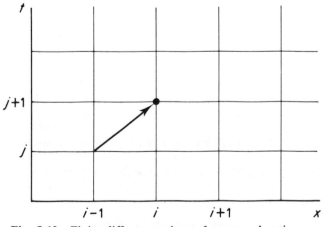

Fig. 5.12 Finite difference scheme for pure advection

Secondly, if $U = 0$ then equation (5.22) is describing dispersion without convection. For this situation, a necessary condition for stability of the numerical solution is that

$$\frac{\delta t}{\delta x^2} E \leqslant \frac{1}{2} \tag{5.25}$$

Assuming that requirement equation (5.24) is satisfied equation (5.22) may be simplified to

$$C_{i,j+1} = C_{i-1,j} + \frac{\delta t}{\delta x^2} E(C_{i-1,j} - 2C_{i,j} + C_{i+1,j}) \tag{5.26}$$

It would now seem possible using equation (5.26) that a program could be written which calculates the unknown concentrations on the $j + 1$ row for all i columns using the known values of C on the j row. Inspection of equation (5.26) shows that if $E = 0$ then criteria equation (5.23) is satisfied for pure convection. However, there is a problem with using equation (5.26) in its existing form.

Figure 5.13 shows a clean river into which effluent is being discharged. For a conservative substance under pure advection the concentration at B at time $j + 1$ must be the same as that at A at time j. Intuitively it can be seen that if dispersion processes are also taking place then the concentration at B must be less than that at A. Equation (5.26) calculates a value for $B(C_{i,j+1})$ which is greater than that at $A(C_{i-1,j})$. This results in a continuous build up of concentration downstream of the discharge point which is artificially too high because of the numerical scheme used.

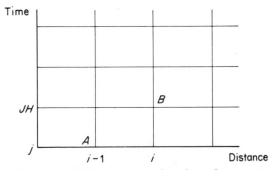

Fig. 5.13   Finite difference scheme for pure advection of a conservative pollutant

A two-step explicit method to overcome this problem is as follows:
The first step is to advect the pollutant downstream for one-time step.
In Fig. 5.14

$$C_{i,n} = C_{i,j-1} \quad \text{for all } i \tag{5.27}$$

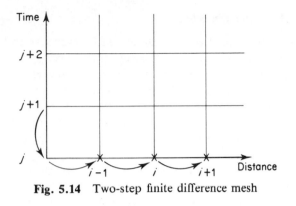

Fig. 5.14 Two-step finite difference mesh

and

$$C_{o,n} = C_{o,j+1}$$

This completes the advection step.

The second step is to calculate new values on the $j+1$ row using only the dispersion and loading terms.

For example, the new concentration at P in Fig. 5.15 is found using the values at X, Y and Z. The reduced equation for this step is

$$\frac{C_{i,j+1} - C_{i,j}}{\delta t} = \frac{E}{\delta x^2} (C_{i-1,j} - 2C_{i,j} + C_{i+1,j}) + La \qquad (5.28)$$

which may be rearranged to give

$$C_{i,j+1} = C_{i,j} + E \frac{\delta t}{\delta x^2} (C_{i-1,j} - 2C_{i,j} + C_{i+1,j}) + \delta t \, La \qquad (5.29)$$

Starting with initial conditions specified on row j at each point i new concentrations may be found on the $j+1$ row for all i. These now become the values used to calculate concentration at $j+2$ etc. and the calculation can continue as long as necessary either to reach steady-state conditions or to display the river quality response characteristics to a time-varying discharge.

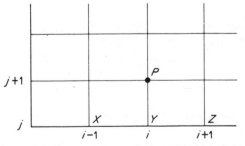

Fig. 5.15 Calculation of pollutant concentration using a two-step finite-difference scheme

5.5 DISSOLVED OXYGEN SAG

Many rivers suffer from depletion of Dissolved Oxygen due to the discharge of organic material which degrades naturally in the river system. The microorganisms which feed on the organic material require oxygen to respire and so reduce the available oxygen in the river. This mechanism is called the Biochemical Oxygen Demand. If the requirements of the pollutant are too large the rate of removal of oxygen is greater than the rate of re-aeration. Oxygen disappears from the river causing distress to other flora and fauna in the area.

Figure 5.16 shows a dissolved oxygen sag curve resulting from the discharge of organic material.

The sag curve is formed because of two competing systems.

Firstly the BOD causing the level to drop and secondly re-aeration which replaces dissolved oxygen from the atmosphere through the surface of the river. The relative magnitude of each will determine the shape of the sag curve. There are other sources and sinks of Dissolved Oxygen, however, BOD is usually the predominant mechanism of depletion.

The sag curve can be predicted for a given effluent discharge using two equations similar to equation (5.21). Assuming the previous river conditions of uniform flow and constant cross-sectional the following equations describe the BOD/DO relationship.

$$\frac{\partial L}{\partial t} = -U \frac{\partial L}{\partial x} + E \frac{\partial^2 L}{\partial x^2} - (K_1 + K_3)L + La \qquad (5.30)$$

$$\frac{\partial C}{\partial t} = -U \frac{\partial c}{\partial x} + E \frac{\partial^2 C}{\partial x^2} - K_1 L + K_2(C_s - C) - B \qquad (5.31)$$

where

    L = first stage BOD concentration
    K_1 = the BOD reaction rate coefficient

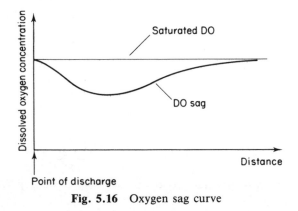

Fig. 5.16   Oxygen sag curve

and the time rate of change in BOD is assumed to be governed by the relationship

$$\frac{dL}{dt} = -K_1 L \tag{5.32}$$

K_3 = the rate coefficient for removal of BOD by sedimentation and adsorption

again this reaction is assumed to be first order of the form

$$\frac{dL}{dt} = -K_3 L \tag{5.33}$$

and

$$\frac{dC}{dt} = K_2(C_s - C) \tag{5.34}$$

K_2 = the re-aeration rate coefficient
C = the dissolved oxygen concentration
C_s = the saturated dissolved oxygen concentration

is assumed to be the equation governing re-aeration.

E = the longitudinal dispersion coefficient
B = the net removal rate of dissolved oxygen for all processes other than biochemical oxidation
La = the rate of addition of BOD along the stretch

In equation (5.30) the rate of removal of BOD is governed by the term $-(K_1 + K_3)L$. *La* is included to account for discharges at different points along the river.

In equation (5.31) $-K_1 L$ governs the rate of removal of dissolved oxygen which is exactly equal to the removal rate for BOD. The term B is included to represent several processes which may affect the sag only to a minor extent. In fact B may be represented by

$$B = B1 + R - P$$

where

$B1$ represents the benthal or bottom deposit demand for oxygen
R represents plant respiration
P represent photosynthesis

These terms may not be included in the initial river model however if the simple model is unable to predict existing conditions it may be necessary to specifically include them in a more sophisticated model.

Equation (5.30) governing the BOD concentration profile may be solved independently of equation (5.31), however, equation (5.31) cannot be solved until the value of L is obtained from equation (5.30). The equations may be solved using the two-step explicit method described earlier by first solving for the BOD concentration and then the DO concentration.

The dispersive and decay sub-equations of equations (5.30) and (5.31) may be written in finite difference form as

$$\frac{L_{i,j+1} - L_{i,j}}{\delta t} = \frac{E}{\delta x^2} (L_{i+1,j} - 2L_{i,j} + L_{i-1,j})$$

$$- \frac{K_1 + K_2}{2} (L_{i,j} + L_{i,j+1}) + La \qquad (5.32)$$

and

$$\frac{C_{i,j+1} - C_{i,j}}{\delta t} = \frac{E}{\delta x^2} (C_{i-1,j} - 2C_{i,j} + C_{i-1,j}) - \frac{K_1}{2} (L_{i,j} + L_{i,j+1})$$

$$+ \frac{K_2}{2} (C_s - C_{i,j} + C_s - C_{i,j+1}) - B \qquad (5.33)$$

collecting terms in equations (5.32) and (5.33) gives

$$L_{i,j+1} \left(1 + \delta t \frac{(K_1 + K_3)}{2} \right) = L_{i-1,j} \frac{\delta t E}{\delta x^2} + L_{i,j} \left(1 - \delta t \frac{(K_1 + K_3)}{2} \right.$$

$$\left. - \frac{2 \, \delta t E}{\delta x^2} \right) + L_{i+1,j} \left(\frac{\delta t E}{\delta x^2} \right) + \delta t La \qquad (5.34)$$

and

$$C_{i,j+1} \left(1 + \frac{\delta t K_2}{2} \right) = C_{i-1,j} \left(\frac{\delta t E}{\delta x^2} \right) + C_{i,j} \left(1 - \frac{\delta t K_2}{2} - \frac{2 \, \delta t E}{\delta x^2} \right)$$

$$+ C_{i+1,j} \left(\frac{\delta t E}{\delta x^2} \right) + \delta t K_2 C_s - \delta t B$$

$$- (L_{i,n} + L_{i,n+1}) \frac{\delta t K_1}{2} \qquad (5.35)$$

A general program for the solution of equations (5.34) and (5.35) is given in Appendix 1.

5.6 LAGRANGIAN MODELS

The finite difference approach uses a fixed Eulerian grid for the solution of the convective diffusion equation. An alternative approach is to use a Lagrangian formulation for the model. This approach assumes the observer is travelling at the same speed as the parcel of water under observation. If it is also assumed that the dispersive mechanism is negligible then biochemical decay is the only source of change within the box. This type of model tracks given parcels of water through time as they move downstream and calculates the rate of biochemical reaction at each time increment. If the position of all the parcels is known at any one time and their concentrations have been calculated, then a one-dimensional plot of concentration against distance can be made for the parameter of interest.

5.6.1 Moving segment model

The basic idea is to simulate the flow in a stream as a series of blocks moving consecutively downstream. Within each block the variations in DO are calculated by summing each hour the changes due to all the processes involved as shown in Fig. 5.17.

The rate of movement of the blocks is related to the flow rate and the cross-sectional area. As the flow varies the cross-sectional area is adjusted and a new time of travel is calculated.

The blocks are considered to be discrete and no exchange of DO or other materials is allowed between blocks. This is not a serious disadvantage because the time step for segmentation is so short that adjacent blocks are unlikely to contain widely differing concentration of oxygen or organic matter.

As the blocks move downstream the rate of oxygen consumption and production will vary due to differences in velocity, nature of the bed, organic concentration, daylight, etc. Some of these differences are associated with the changing physical characteristics of the river and for this reason the model river is divided into reaches. Within each reach the rates of re-aeration and benthal respiration are regarded as uniform and the biomass of the rooted flora is considered to be evenly distributed.

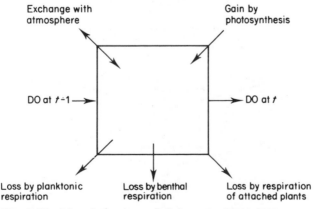

Fig. 5.17 Mass balance on DO in a moving segment

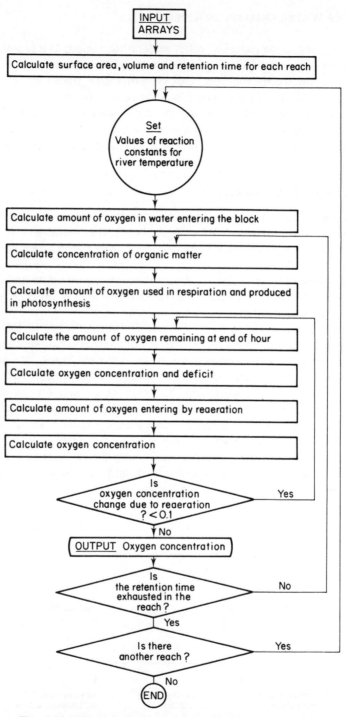

Fig. 5.18 Flow diagram of the dissolved oxygen model

The end of a reach marks a point of discontinuity in the physical conditions or may mark a discontinuity in the oxygen profile due to a weir. Sub-division of the river into reaches is also made at places where any major tributary or effluent enter.

The mathematical representation of the oxygen concentration in a block may be summarized as follows

$$\frac{dC}{dt} = P \pm RA - R \tag{5.36}$$

where

$$P = \frac{B1 \times Pmx}{K} \times \frac{I \times SA}{IK} + B2 \times Pmx \times \frac{I}{Ik} \tag{5.37}$$

$$RA = K2(C_s - C) \tag{5.38}$$

$$RE = K1(\text{BOD}) \times SA \tag{5.39}$$

and

$$Ct = Co + \int_0^t \frac{dc}{dt}$$

P = photosynthesis
RA = re-aeration
RE = respiration
I = light intensity
Ik = light intensity corresponding to Pmx
SA = surface area
B_1 = biomass of attached plants and algae
B_2 = biomass of planktonic algae
K = extinction coefficient for light
K_3 = respiration rate of bottom deposits
Pmx = maximum rate of photosynthesis

The inputs to the model define the upstream boundary conditions, the values of all the rate coefficients for each reach and the climatic conditions in terms of light intesity and temperature. Once one block has been followed through all reaches then the program loops back to follow the oxygen concentration in the next block not shown in Fig. 5.18. A listing of the program for this model written in FORTRAN is given in Appendix 2 of this chapter.

5.7 OPERATIONAL MODEL

Dispersion and moving segment models are generally used as planning tools since they predict the average dissolved oxygen level for a particular set of conditions. There are some circumstances in polluted rivers where an operational model is required to give much more detailed and accurate predictions.

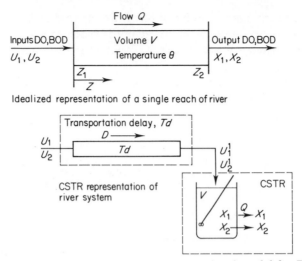

Fig. 5.19   Diagrammatic representation of an operational model for DO in a stream

This involves real time simulation and monitoring usually in connection with control of a river water abstraction or an in-river aeration system.

Such a model can be constructed by dividing the river into a series of reaches, each of which is represented as a stirred tank reactor with reaction and time delay, as shown in Fig. 5.19. Where the biochemical oxygen demand and the reaction are the main sink and source, the equation for the DO is as follows:

$$V \frac{dC2}{dt} = Q(C1_{t-d} - C2_t) + VK2(Cs - C2_t) - VK1L2_t \qquad (5.40)$$

Rate of change at end of reach = Flux in − Flux out + Rate of re-aeration
− Rate of deoxygenation

The time delay in the reach is variable depending upon fresh water flow Q and the volume of the reach V.

The response of the system is illustrated by comparing the observed and predicted data for a pulse input (see Fig. 5.20).

In such a model unaccounted variables are lumped together into a stochastic input $R_{t,2}$. There are also stochastic variations in the output due to errors in measurement.

$$C''_{0,t} = f(C_{0,t}),\ N(t,\ 2) \qquad (5.41)$$

Hence the full model for the reach can be written as

$$V \frac{dC2}{dt} = Q(C1_{t-d} - C2_t) + K2(Cs - C2_t) - K1L2 + R_{t,2} \qquad (5.42)$$

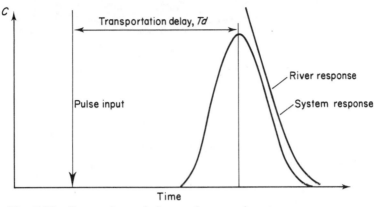

Fig. 5.20 Comparison of river and assumed system response

where

$$C2_t = C1_{2,t} \pm N(t)$$

As $(t, 2)$ and (t) have different frequencies they can be separated by time series analysis. For example, if $R(t, z)$ is a low-frequency stochastic term due to the omission of some slowly varying sink and $N(t, 2)$ is a random variable applied to each DO measurement which are made at frequent intervals then separation could be achieved by time averaging over a number of DO observations, since $R(t, 2) = 0$. In practice much more complex statistical techniques are employed notably the Kalman filter.

Since $N(t, z)$ and $R(t, z)$ can be separated the state parameter $K1$ and $K2$ can be evaluated and constantly updated by recursive estimates on observed data.

5.8 OPTIMIZATION MODEL

River models based on dispersion or moving segments can be used to generate a matrix of transfer coefficients by running the model several times with unit changes in BOD loadings. Thus, for a simple situation with two reaches and two waste treatment plants as shown in Fig. 5.21 the DO/BOD relationship may be represented by a 2×2 matrix of coefficients. Each transfer coefficient ϕ represents the change in DO in a reach i for unit change in BOD loading from treatment plant j.

$$\begin{bmatrix} \phi_{11} & \phi_{12} \\ \phi_{21} & \phi_{22} \end{bmatrix}$$

Given the data on the cost of treatment at the two plants $C1$ and $C2$ (units costs per kg BOD removed per day) and the improvement in DO required $D1$

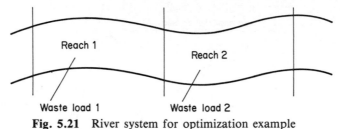

Fig. 5.21   River system for optimization example

and $D2$ (gO_2 per m^3) then the problem can be solved by a suitable optimization routine. The objective function can be written as

   Minimize

$$C = \sum_{j=1}^{n} C_j W_j$$

Where

   $C = $ Total cost

   $W_j = $ Number of kg of BOD removal per day at treatment plant j

   $C_j = $ Cost per kg of BOD removed per day at treatment plant j

and the constraint equation becomes

$$W1 \times \phi_{11} + W2 \times \phi_{21} \geqslant D1$$

$$W2 \times \phi_{12} + W2 \times \phi_{22} \geqslant D2$$

giving the following data

$$D1 = 1$$

$$D2 = 2$$

$$W1 = 10$$

$$W2 = 20$$

$$\begin{bmatrix} \phi_{11} & \phi_{12} \\ \phi_{21} & \phi_{22} \end{bmatrix} = \begin{bmatrix} 0.01 & 0.001 \\ 0.005 & 0.01 \end{bmatrix}$$

the problem may be solved by the simplex method.

	$W1$	$W1$	
$R1$	0.01	0.001	1
$R2$	0.005	0.01	2
$-C$	10	20	0

The pivot element must lie in $W1$ column and the ratios $2/0.005$ and $1/0.01$ show that it must be the element in the top row. The tableau can then be transformed using the method outlined in Chapter 1. The new tableau therefore becomes

	$R1$	$W2$	
$W1$	100	0.1	100
$R2$	-0.5	0.09	1.5
$-C$	-1000	10	-000

In this case the pivot element must lie in the $W2$ column and considering the constraint ratios:

$$\frac{1.5}{0.09} = 166$$

$$\frac{100}{0.1} = 1000$$

it must be element $R2/W2$. Transforming the tableau about that element gives

	$R1$	$R2$	
$W1$	105.3	10.5	85
$W2$	-52.5	105	157.5
$-C$	0	-11445	-4000

which is the final solution indicating that the improvement in DO can be obtained by an additional 85 Kg per day removal at treatment plant 1 and an additional 157.5 Kg per day removal at treatment plant 2. The cost will be 4000 and the improvement in DO will be

Reach 1 $85 \times 0.01 + 157.5 \times 0.001 = 1 \, \text{mg} \, 1^{-1}$
Reach 2 $85 \times 0.01 + 157.5 \times 0.005 = 2 \, \text{mg} \, 1^{-1}$

5.9 MODELS OF DISCHARGE

A model for a point discharge has already been described in Chapter 1. As shown in Fig. 1.1, the concentration of a conservative pollutant at the end of the mixing zone can be calculated from a simple mass balance:

$$Cm = \frac{Q1 \times C1 + Q2 \times C2}{Q1 + Q2}$$

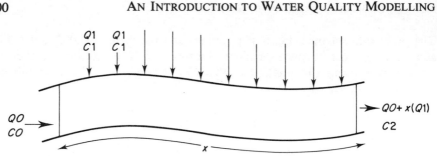

Fig. 5.22   Conceptual model of a diffuse source

where

$$Cm = \text{concentration at end of mixing zone}$$
$$C1 \text{ and } Q1 = \text{concentration and flow in fresh water}$$
$$C2 \text{ and } Q2 = \text{concentration and flow in effluent}$$

Where the discharge is from a diffuse source the representation can be made by dividing the river into reaches so that within each reach the discharge is uniform along its length (see Fig. 5.22)
the mass balance on the conservative pollutant is given by

$$\frac{d}{dx}(QO + Q1x)C2)) = Q1C1$$

where

$$QO \text{ and } CO \text{ are flow and concentration upstream}$$
$$x \text{ is the length of the reach}$$
$$Q1 \text{ is additional run-off per unit length}$$
$$C1 \text{ is the concentration in the run-off}$$

this integrates to give

$$C2 = COF + C1(1 - F)$$

where

$$F = \text{Dilution Factor}$$

$$= \frac{QO}{QO + Q1X}$$

As with point discharges it is assumed that the pollutant mixes completely with the river water before the end of the reach.

For non-conservative substances, the same approach may be used suitably modified to take account of the chemical and/or biological processes that alter the concentration of the substance concerned.

For DO, the appropriate equation is as follows:

$$\frac{d}{dx}(Q + qx)d = qr_D + K_2A(d_s - d) - K_1Az - K_3Aa$$

which may be integrated to give

$$d(L) = d_0 F^\delta + \frac{1}{q\delta} r_D q + K_2 A d_s - \frac{K_1 A r_2}{\alpha} - \lambda K_3 A r_A$$

$$(1 - F^\beta) - \frac{K_1 A}{q(\delta - \alpha)} z_0 - \frac{r_2}{\alpha}(F^\alpha - F^\delta)$$

$$- \frac{\lambda K_3 A}{q(\delta - \beta)} a_0 - \frac{r_A}{\beta}(F^\beta - F^\delta)$$

where

$$\delta = 1 + \frac{K_2 A}{q} \qquad \alpha = \frac{1 + (K_1 + K_5)A}{q} \quad \text{and} \quad \beta = \frac{1 + (K_3 + K_4)A}{q}$$

Similarly, for ammonia the equations may be given as:

$$\frac{d}{dx}(Q + qx)a = qr_A - (K_3 + K_4)A_a$$

and

$$a(L) = a_0 F^\beta + \frac{r_A}{\beta}(1 - F^\beta)$$

and the other symbols are defined as follows:

A = average cross-sectional area
a = concentration of ammoniacal nitrogen
d = concentration of dissolved oxygen
λ = relative oxygen usage in nitrification
r_D = DO in run-off
r_A = Ammoniacal nitrogen concentration in run-off
r_z = BOD in run-off
d_s = DO saturation
K_1 = BOD decay coefficient
K_2 = reaeration coefficient
K_3 = nitrification coefficient
K_4 = volatilization coefficient for ammonia
K_5 = BOD sedimentation coefficient
q = $Q1$ as defined in Fig. 5.22

5.10 BIBLIOGRAPHY

Modelling of dissolved oxygen in rivers has received more attention than other aspects of water quality and the literature is therefore more voluminous. The following suggestions are examples of the different approaches that have been used.

Rinaldi, S., Soncini-Sessa, R., Stehfest, H. and Tamara, H. (1979). *Modelling and Control of River Water Quality.* McGraw-Hill, New York.

Beck, M. B. (1978). 'Modelling of Dissolved Oxygen in a Non-Tidal Stream', Chapter 6. In: *Mathematical Models in Water Pollution Control* (ed. James, A.). John Wiley & Sons, Chichester.

Velz, C. J. (1970). *Applied Stream Sanitation.* John Wiley & Sons, New York.

Dresnack R. and Dubbins, V. E. (1968). Numerical analysis of BOD–DO profiles, *American Society of Civil Engineers and Journal of Sanitary Engineering Div.* **8A5**, 789–807.

Owens, M. (1969). Measurement of primary productivity in flowing waters. In: *Measurement of Primary Productivity.* IBP Handbook No. 6 (ed.) Wollenweider. Blackwood, Edinburgh.

Linsley, R. K. and Franzini, J. B. (1979). *Water Resources Engineering.* McGraw-Hill, New York.

APPENDIX 1

Listing of dispersion model—(BASIC)

```
100   REM FRESH WATER MODEL
110   REM  – – – – – – – – – – – – –
120   REM FINITE DIFFERENCE APPROACH
130   REM  – – – – – – – – – – – – – – – – – – – –
140   REM SIMPLE EXPLICIT SOLUTION
150   REM  – – – – – – – – – – – – – – – – –
160   REM UNITS USED
170   REM  – – – – – – –
180   REM
190   REM DISTANCE METERS
200   REM TIME SECS
210   REM VELOCITY METRES/SEC
220   REM REACTION RATES / SEC
230   REM
240   REM SUGGESTED VALUES FOR TRIAL RUN
250   REM  – – – – – – – – – – – – – – – – – – – – –
260   REM SATOX 8.5
270   REM STREAM WIDTH 12.5
280   REM STREAM DEPTH 0.6
```

```
290   REM K1 0.00002 K3 0.000005
300   REM DISPERSION COEFF 10
310   REM VELOCITY 0.2
320   REM TIME STEP 2000
330   REM MESH DISTANCE 400
340   REM BOD LOAD 10000 – GRAMS
350   OPEN 1,3
360   DIM LO(50), CO(50), L1(50), C1(50)
370   DIM S(50)
380   PRINT 1, "INPUT SATOX CONC"
390   INPUT C9
400   PRINT 1, "INPUT STREAM WIDTH"
410   INPUT B
420   PRINT 1, "INPUT STREAM DEPTH"
430   INPUT D
440   PRINT 1, "INPUT K1, K3
450   INPUT K1, K3
460   PRINT 1, "INPUT DISP COEFF, VELOCITY"
470   INPUT D1,U
480   PRINT 1, "INPUT TIME STEP, DELTA X"
490   INPUT H,X
500   FOR I = 1 TO 50
510   LO(I) = 0
520   CO(I) = C9
530   S(I) = 0
540   NEXT I
550   PRINT 1, "INPUT NO. OF DISCHARGES"
560   INPUT I1
570   FOR I = 1 TO I1
580   PRINT 1, "INPUT LOAD AND MESH NO."
590   INPUT W,N
600   S(N) = W/(B*D*X)
610   NEXT I
620   T = TP – 1
630   PRINT 1, "INPUT NO OF TIME INCS"
640   INPUT N9
650   PRINT 1, "INPUT PRINT STEP"
660   INPUT TP
670   T = TP = 1
680   FOR N = 1 TO N9
690   T = T + 1
700   PRINT 1, "TIME STEP";N
710   PRINT    " – – – – – – – "        1,
```

```
720   FOR I = 1 TO 47
730   II = 48
740   CO(II) = CO(II − 1)
750   LO(II) = LO(II − 1)
755   II = II − 1
760   NEXT I
770   FOR I = 3TO 47
780   K2 = 0.01*U^0.5/D^1.5
790   Z1 = (1/(1 + H*(k1 + K3)/2))
800   Z2 = LO(I − 1)*H*D1/X^2
810   Z3 = L0(I)*(1 − H(K1 + K3)/2 − 2*H*D1/X^2)
820   Z4 = LO(I + 1)*H*D1/X^2
830   L1(I) = Z1*(Z2 + Z3 + Z4 + S(I))
840   Y1 = (1/(1 + H*K2/2))
850   Y2 = CO(I − 1)*H*D1*X^2
860   Y3 − CO(I)*(1 − H*K2/2 − 2*H*D1/X^2)
870   Y4 = CO(I + 1)*H*D1/X^2
880   Y5 = H*K2*C9
890   Y6 = (LO(I) + L1(I))*H*K1/2
900   C1(I) = Y1*(Y2 + Y3 + Y4 + Y5 − Y6)
910   IF T< >TP THEN940
920   PRINT 1, "MESH NO";I
930   PRINT 1, "BOD = ";L1(I);"D.O. = ";C1(I)
940   NEXT I
950   FOR J = 1 TO 50
960   LO(J) = L1(J)
970   CO(J) = C1(J)
980   NEXT J
990   IF T< >TP THEN1010
990   IF T< >TP THEN1010
1000  T = 0
1010   CO(1) = 0
1020  CO(2) = 0
1030  CO(49) = 0
1040  CO(50) = 0
1050  LO(1) = 0
1060  LO(2) = 0
1070  LO(49) = 0
1080  LO(50) = 0
1090  NEXT N
1100  END
```

PPENDIX 2

Moving block DO model—(FORTRAN)

```
1           DIMENSION FLOWF(N,365),SOLRAD(365,24),FLOWW(N,365),TEMPR(N,365)
2          &DOF(N,365,24),DOW(N,365,24),BCDF(N,365,24)BODW(N,365,24)
3           PHOT(N),        STORE(N),BIOMAT(N)
4          &,REAER(N),GRAF(500,50),DEOXR(N),BENR(N),BIOMR(N)
5           REAL KK,K1(5),K9(5),LEN(5),K2,K3,K4(N),K5(N),K6(N),K7(N),K8(N)
6           READ(5,119) IY
7     119   FORMAT(I1)
8           DO 16 I = 1,5
9           DO 18 K = 1,10
10          FLOWF(2,K) = FLOWF(2,K) + FLOWF(1,K)
11          FLOWW(2,K) = FLOWW(2,K) + FLOWW(1,K)
12          FLOWF(3,K) = FLOWF(3,K) + FLOWF(2,K)
13          FLOWW(3,K) = FLOWW(3,K) + FLOWW(2,K)
14          FLOWF(4,K) = FLOWF(4,K) + FLOWF(3,K)
15          FLOWW(4,K) = FLOWW(4,K) + FLOWW(3,K)
16          FLOWF(5,K) = FLOWF(5,K) + FLOWF(4,K)
17    18    FLOWW(5,K) = FLOWW(5,K) + FLOWW(4,K)
18          DO 35 J = 1,N
19          DO 37 L = 1,N
20          VOLR(N) = ((FLOWF(N,K) + FLOWW(N,K) - (FLOWF(N - 1,K) + FLOWW(N - 1,K))*
21                  (CA(N))*(LENR(N))
22          WIDTHR(N) = VOLR(N)/FLOWF(N,K) + FLOWW(N,K))
23    37    SURR(N) = LENR(N)*WIDTHR(N)
24          IF(IY.NE.1) GO TO 50
25          WRITE(6,333)
26          WRITE(6,334) J
27          WRITE(6,335)
28          WRITE(6,336)
29    333   FORMAT('1',57X, '*****')
30    334   FORMAT(58X, '* DAY ',13,' *')
31    335   FORMAT(58X, '*****',////)
32    336   FORMAT('******************************'
33         &'*************************************'
34         &'************************************')
35    50    DO 23 N = 1,5
36          K5(N) = REAER(N)*1.047**(TEMPR(N,J) - 20.0)
37          K6(N) = DEOXR(N)*1.047**(TEMPR(N,J) - 20.0)
38          K7(N) = BENR(N)*1.047**(TEMPR(N,J) - 20.0)
39          K8(N) = BIOMR(N)*1.047**(TEMPR(N,J) - 20.0)
40          K9(N) = PHOT(N)*1.047**(TEMPR(N,J) - 20.0)
41          RETR(N) = VOLR(N)/(FLOWF(N,J) + FLOWW(N,J)
42          IF(IY.NE.L) GO TO 23
43          WRITE(6,601)N,RETR(N),K5(N),K6(N),K8(N),K9(N)
44    601   FORMAT(////,'REACH',12,6X'RET. TIME = ',F.7.1,6X,'K5 TO K9 '
45         &'VALUES... ',5F10.4)
46    23    CONTINUE
47          DO 34 K = 1,24
48            CCCCCCCCCCCCCCCCCCCCCCCCCCCCCCCCCCCCCCCCCCCCCCCCCCCCCCCCCCCCCC
```

```
49      C
50      C  THIS IS THE START OF THE STORY OF A BLOCK OF WATER, ORIGINALI
51      C  AT THE BEGINNING OF REACH 1.   24 BLOCKS START THE JOURNEY
        EACH
52      C  DAY.                  NOTE THAT AT ANY GIVEN TIME, MANY BLOCKS AF
        TRAVELLING.
53      C
54      CCCCCCCCCCCCCCCCCCCCCCCCCCCCCCCCCCCCCCCCCCCCCCCCCCCCCCCCCCCCCC
55          L = K − 1
56      CCCCCCCCCCCCCCCCCCCCCCCCCCCCCCCCCCCCCCCCCCCCCCCCCCCCCCCCCCCCCC
57      C
58      C  FOR THIS BLOCK, L IS ORIGINALLY SET TO K WHICH IS THE HOUR
59      C  OF THE DAY THAT THE BLOCK STARTS FROM AND IS INCREMENTED B
60      C  1 UNTIL IT REACHES 24, AFTER WHICH IT IS SET TO 1 AND "KI"
61      C  IS INCREMENTED. THIS CARRIED ON UNTIL THE BLOCK REACHES THE
62      C  END OF THE RIVER.
63      CCCCCCCCCCCCCCCCCCCCCCCCCCCCCCCCCCCCCCCCCCCCCCCCCCCCCCCCCCCCCC
64          I = 0
65          KI + J
66      21     I = I = 1
67      IF(I.NE.2) GO TO 97
68      R = 1.25
69      STORE(2) = OXSAT − (OXSAT − OXCON)/R
70   97 IF(IY.NE.1) GO TO 83
71      IF(I.EQ.1) WRITE(6,889)
72  889 FORMAT(//,'*****************************'  ,/,
73      &'* NEXT BLOCK ENTERS RIVER HERE *',/,
74      &'*****************************************',//)
75   83 IF(I.GT.5) GO TO 33
76      IF(I.EQ.1) IBASE = 0
77      IF(I.EQ.2) IBASE = 16
78      IF(I.EQ.3) IBASE = 25
79      IF(I.EQ.4) IBASE = 33
80      IF(I.EQ.5) IBASE = 42
81      IF(IY.NE.I) GO TO 51
82      WRITE(6,336)
83      WRITE(6,445) I
84      WRITE(6,336)
85  445 FORMAT(50X, 'BLOCK HAS JUST ENTERED REACH',I2)
86  CCCCCCCCCCCCCCCCCCCCCCCCCCCCCCCCCCCCCCCCCCCCCCCCCCCCCCCCCCCCCC
87  C  I IS INCREMENTED EACH TIME A BLOCK CHANGES REACHES SO THAT IT
88  C  CAN USE DATA FOR THE CORRECT REACH THROUGHOUT ITS JOURNEY,
89  C  WHEN THE BLOCK HAS REACHED THE END OF THE RIVER, THE OXYGEN
90  C  CONCENTRATION AT THE END OF EACH REACH FOR THAT BLOCK IS OUT-
91  C  PUT TOGETHER WITH THE STARTING HOUR AND DAY FOR THAT BLOCK.
92  C
93  CCCCCCCCCCCCCCCCCCCCCCCCCCCCCCCCCCCCCCCCCCCCCCCCCCCCCCCCCCCCCC
94   51 KNAPPI = RETR(I) + 1.
95      RESP = 0
96      DO 59 IPP = 1,KNAPPI
97      IPOS = IBASE + IPP
```

```
 8          L = L + 1
 9     CCCCCCCCCCCCCCCCCCCCCCCCCCCCCCCCCCCCCCCCCCCCCCCCCCCCCCCCCCCCCCCC
 0     C
 1     C   A IS SET TO 1 IN THIS LOOP EXCEPT FOR THE LAST CYCLE WHEN IT
 2     C   IS SET TO THE FRACTION OF THE HOUR THAT THE BLOCK REMAINS IN
 3     C   THIS REACH BEFORE PASSING ON TO THE NEXT ONE.
 4     C
 5     CCCCCCCCCCCCCCCCCCCCCCCCCCCCCCCCCCCCCCCCCCCCCCCCCCCCCCCCCCCCCCCC
 6          IF(L.GT.24) KI = KI + 1
 7          IF(L.GT.24) L = 1
 8          ITIME = (KI − 1)*24 + L
 9          IF(I.EQ.1.AND.IPP.EQ.1) OXCON =
 0         &(DOF(I,KI,L)*FLOWF(I,KI) + DOW(I,KI,L)*
 1         &FLOWW(I,KI))/FLOWF(I,KI) + FLOWW(I,KI))
 2     15   AMNTR = (FLOWF(I,KI) + FLOWW(I,KI))*OXCON
 3    500   AMNTB = SURR(I)/RETR(I)*K7(I)
 4    501   IF(IPP.EQ.1.AND.I.NE.1) ORGC = (BODF(I,KI,L)*FLOWF(I,KI)
 5         & + BODW(I,KI?L)*FLOWW(I,KI))/(FLOWW(I,KI) + FLOWF(I,KI)) + ORGC
 6    502   IF(IPP.EQ.1.AND.I.EQ.1) ORGC = (BODF(I,KI,L)*FLOWF(I,KI) + BODW(I,KI,L)
 7         &)*FLOWW(I,KI))/(FLOWF(I,KI) + FLOWW/I,KI))
 8    503   IF(IPP.NE.1) ORCG = ORGC − RESP/(FLOWF(I,KI) + FLOWW(I,KI))
 9          IF(ORGC.LT.0) ORGC = 0.0
 0    504   KK = K6(I)
 1    505   AMNTP = ORGC*(EXP(− KK))*(FLOWF(I,KI) + FLOWW(I,KI))
 2          RESP = RESP + AMNTP
 3    506   AMNTAI = BIOMAT(I)*K8(I)*(1.0/RETR(I))
 4    507   AMPHI = K9(I)*(BIOMAT(I)/RETR(I))*(0.5*SOLRAD(KI,L)*
 5         &(70. − SOLRAD(KI,L))/70.)
 6          AMREA = 0
 7          OXCON = 20000
 8          OXSAT = 9
 9     31   OXAM = (AMNTR + AMREA − (AMNTB + AMNTP + AMNTAI) + AMPHI)
 0          TEMP = OXCON
 1          OXCON = OXAM/((FLOWF(I,KI) + FLOWW(I,KI)))
 2          KK = K5(I)
 3          AMREA = KK/100.*(OXSAT − OXCON)*(SURR(I)/RETR(I))
 4          IF(ABS(TEMP − OXCON).GT.0.2) GO TO 31
 5          IF(IY.NE.1) GO TO 67
 6          WRITE(6,125)K,J,I,L,KI,AMNTR,AMNTB,ORGC,AMNTP,AMNTAI,AMPHI,OXCON
 7    125   FORMAT(//,'BLOCK THAT STARTED AT ',I2,' 0''CLOCK ON DAY',
 8         &I3, 'IS NOW IN REACH ',I1, 'AT ',I3,' 0''CLOCK ON DAY ',I3,
 9         &/, 'AMNTR = ',F11.0,' AMNTB = ',F5.0,' ORGC = ',F4.1,
 0         &' AMNTP = ',F8.0,' AMNTAI = ',F6.0,' AMPHI = ',F6.0,
 1         &' OXCON = ',F6.2)
 2     67   IF(ITIME.GT.500.OR.IPOS.GT.50) GO TO 59
 3          GRAF(ITIME,IPOS) = OXCON
 4     59   IF(I.EQ.1) STORE(I) = OXCON
 5          IF(I.NE.1) STORE(I + 1) = OXCON
 6          GO TO 21
 7     33   WRITE(6,100) K,(STORE(I),I = 1,6),K
 8     CCCCCCCCCCCCCCCCCCCCCCCCCCCCCCCCCCCCCCCCCCCCCCCCCCCCCCCCCCCCCCCC
```

```
149   C   NOW A BLOCK IS AT THE END OF THE RIVER AND SO ITS RESULTS WERE
150   C   JUST OUTPUT FOR THE END OF EACH REACH TOGETHER WITH ITS START-
151   C   ING TIME AND DATE. WE THEN GO BACK IN TIME TO FOLLOW THE NEXT
152   C   BLOCK DOWNSTREAM.    CCCCCCCCCCCCCCCCCCCCCCCCCCCCCCCCCCCCC
153   34  CONTINUE
154       WRITE(6,336)
155   35  CONTINUE
156       DO 717 I = 1,46
157       WRITE(6,921) I
158   921 FORMAT(//,30X,'*** POSITION ',12,' ***')
159   717 WRITE(6,923)(GRAF(J,I),J = 1,198)
160   923 FORMAT(12F10.1)
161   100 FORMAT('*',19X,I8,6F11.4,17,29X'*',/)
162       IF(IY.EQ.2) CALL PLOT(GRAF)
163       END
```

NOTES ON MOVING BLOCK MODEL

This is written in FORTRAN, hence the need to declare all the variables as REAL i.e. can take decimal values or INT i.e. integers only. Also the need for FORMAT statements on input and ouput.

The variables are as follows:

FLOWF	= Flow of freshwater
FLOWW	= Flow of waste
SOLRAD	= Solar radiation
TEMPR	= Temperature of reach
DOF	= Dissolved oxygen in freshwater
DOW	= Dissolved oxygen in waste
BODF	= BOD of freshwater
BODW	= BOD of waste
PHOT	= Photosynthetic coefficient (K9)
BIOMAT	= Biomass of benthic plants
REAER	= Coefficient of reaeration (K5)
RETR	= Retention time in reach (see line 41)
SURR	= Surface area of reach
WIDTHR	= Width of reach

Chapter 6

Models of Water Quality in Estuaries

D. J. ELLIOTT AND A. JAMES

6.1 INTRODUCTION

Conditions in estuaries are more complex than in rivers and are therefore more difficult to model. Hydraulically there are three complications:

(a) Two directional flow resulting from the interaction between freshwater flow and tidal movements.
(b) Density differences between the freshwater (also effluents) entering an estuary and the saline water entering from the sea. This results in a variable degree of mixing. Also the density differences tend to create a gravitational circulation pattern.
(c) Rivers tend to widen out as they enter estuaries and the resulting water body is often subject to a horizontal circulation pattern due to the Coriolis effect.

Biologically, estuaries are a distinct part of the river system. The transition zone between freshwater and marine conditions is unfavourable for the vast majority of aquatic plants, animals and micro-organisms. The central section of estuaries with transitional and constantly varying salinities is biologically a desert inhabited only by phytoplankton or by those organisms like burrowing worms or bivalve molasses which can to some extent control the salinity around themselves.

From a modelling viewpoint it is important to identify the populations at risk so that appropriate environmental parameters can be used. The standards generally adopted for estuaries are as follows:

(i) No nuisance standard—Dissolved oxygen $\not< 1$ mg 1^{-1} and no surface slick of oil and a limit on suspended solids.
(ii) Coarse fish standard—Dissolved oxygen $\not< 3$ mg 1^{-1} and a limit on toxins.

The common feature of all these standards is the concentration of dissolved oxygen. Estuary models, like river models therefore tend to be concerned with BOD/DO relationships. There is an important difference however in the

109

method of expressing the oxygen demand of the organic matter. Retention times in estuaries are much greater than in the freshwater reaches of rivers and allow time for complete oxidation to occur. The concept of BOD (5-day retention) is therefore replaced by UOD (20-day retention).

6.2 ESTUARINE HYDRAULICS

Water movements in esturaries may be represented by the diagram shown in Fig. 6.1. If no density differences occur, particles at AA at high tide would move down the estuary and return to $A'A'$ at the next high tide.

The net movement downstream is due to advection by freshwater flow. In the region of density differences, particles at AA would be distributed over the line BB after one tidal cycle. It can be seen from Fig. 6.1 that increasing the freshwater flow does not directly increase the flushing effect at all depths. A rotational movement occurs with significant net upstream movement near the estuary bed. These show that even under flood flow conditions a net upstream movement of water (and associated solutes) near the bed. There is a tendency for the less dense freshwater to flow over the top of the estuarine water.

Near the head of the estuary the current pattern changes as water and solutes in the lower layers become incorporated in the upper layers and travel downstream.

The overall tidal average pattern of water movement is shown in Fig. 6.2. This describes the movement of materials (such as pollutants) due to advection, i.e. transport caused by the bulk movement of the water. Pollutants also move due to non-advective processes, particularly eddy diffusion. Estuaries may therefore be represented by the three-dimensional version of the convective-diffusion equation. But the complexity of model formulation and the major task of data collection have forced most modellers to simplify into one or at the most two dimensions.

The decision on the number of dimensions is based upon the degree of

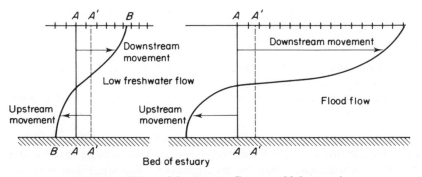

Fig. 6.1 Effect of freshwater flow on tidal excursion

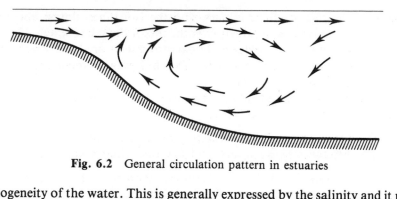

Fig. 6.2   General circulation pattern in estuaries

homogeneity of the water. This is generally expressed by the salinity and it may
be used as a basis for estuarine classification as shown in Fig. 6.3.

In Fig. 6.3a the isohalines are almost vertical showing that water quality
does not vary with depth. Provided that the estuary is laterally homogeneous
a 1-*d* longitudinal model would be sufficient to represent the quality pattern.

In Fig. 6.3b the isohalines are diagonal showing some degree of stratification. Estuaries of this type are probably best represented by 2-d models or by
layered 1-d models.

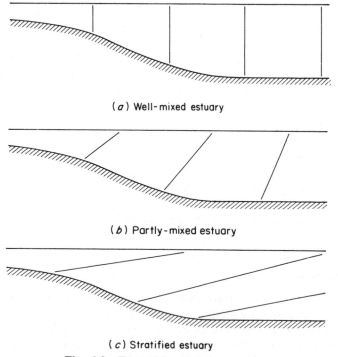

(*a*) Well-mixed estuary

(*b*) Partly-mixed estuary

(*c*) Stratified estuary

Fig. 6.3   Types of mixing in estuaries

Fig. 6.3c shows a stratified estuary with the isohalines almost horizontal. The freshwater movement is largely over the top of the saline water. The latter enters the estuary as a salt-wedge and tidal movements are largely represented by the advance and retreat of this wedge. Depth-averaged models are not really suitable for this type of estuary.

In the absence of detailed salinity data, it is possible to estimate the degree of mixing from the D/T rates where

D = volume of freshwater entering the estuary during a tidal cycle
T = difference between high and low tide water volume

Where the D/T ratio is low (i.e. < 0.1) the estuary tends to be well mixed. D/T ratios of 0.1 to 1.0 are characteristic of partly stratified conditions and ratios over 1.0 signify the salt-wedge type of estuary.

6.3 ESTUARINE MODELS

Three techniques have been used with reasonable success in estuary models, analytical, numerical or simulation. Other techniques such as stochastic or correlation models have limited applicability and will not be discussed further.

Analytical methods usually involve a simplified form of the convective diffusion equations and have been used in the solution of general problems in homogeneous tidal zones. They are limited in application and produce solutions similar to those obtained for rivers.

Simulation models have mainly been used in both homogeneous and non-homogeneous estuarine zones for one and two dimensional applications.

Many models are based on the numerical solution of the convective diffusion equation in one or two dimensions.

As with freshwater river models, it is assumed that the discharges of pollutant mix rapidly with the surrounding environment so that the combined flow is neutrally buoyant and has negligible effect on the existing flow patterns.

If density differences are significant, as in the case of cooling-water problems for power stations, then it may be necessary to simultaneously solve the momentum equations and the continuity equations of the flow field. This sort of problem is difficult to solve numerically and hydraulic studies may be preferred.

6.3.1 Time scale

There are three general categories of estuary model based on the time scale used. Firstly, steady-state models assume that the longitudinal distribution of any water-quality parameter is identical at corresponding points and lines over successive tidal cycles. These models are often one-dimensional using tidal average values of parameters for depicting the half-tide situation. Advection is caused only by freshwater flow since the tidal motion averages to zero over one tidal cycle. This type of model is generally applied to freshwater zones of estuaries or to estuaries which are vertically well mixed. They are used to predict conditions during which the tidal range does not vary greatly from tide to tide.

The second model category is that in which parameters are allowed to vary from one tidal cycle to the next. Conditions within the tidal cycle are not reproduced and advective motion in attributed to freshwater flow. These models are used in situations where intermittent injection of pollutants occur or where continuous injections which vary slowly with a time scale larger than a tidal cycle.

The third category are general time varying models reproducing conditions within the tidal cycle. Advective motion is represented by the instantaneous resultant of tidal and freshwater flows. Such models are used in estuaries subject to significantly varying discharges of pollutants and freshwater flow.

6.3.2 Model use

As mentioned above, different categories of model have been used to describe different estuary situations. However, most models have been produced to predict the effect of new or altered combinations of discharge on estuary water quality. The conditions most likely to produce the poorest water quality are used in the model. That is summer conditions of low freshwater flow and high temperatures which may last for a period of several weeks. This may be considered to be a pseudo-steady-state situation for continuous effluent discharges and so steady-state models have been found to be useful. Even in cases where dynamic models are available they have been run for sufficient length of time to produce the steady-state 'worst' condition.

6.3.3 Simulation models

These represent the traditional form of solution for estuary water-quality problems in which the advective and dispersive processes are replaced by a single mixing operation. This type of model assumes that the estuary is divided longitudinally into a number of segments. Each segment is assumed to be

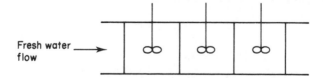

Fig. 6.4 Diagramatic representation of an estuary as a series of stirred tanks

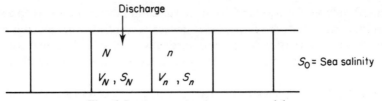

Fig. 6.5   A segmented-estuary model

uniformly well mixed so that the concentration in any segment may be represented by a single statistic.

In this way the estuary can be visualized as consisting of a series of continuously stirred tanks. Several models of this type have been developed but only two will be described here. The first is a simple model which can be used without the aid of digital computers and the second is a model which was used as a basis for the Thames study.

In the first of these models, the estuary is divided into a number of segments of length equal to half the mean tidal excursion. The effluent is assumed to be instantly mixed in the outfall segment and have the same characteristics as the freshwater flow. Advection and longitudinal mixing carry the effluent to adjacent segments downstream while upstream segments are affected only by longitudinal mixing. It is necessary to calculate the retention period of each segment which is equal to that of the freshwater flow on the assumption that the added effluent does not significantly affect the mixing patterns.

Consider any segment n of total midtide volume V_n and salinity S_n.

If the seawater salinity $= S_0$ and freshwater salinity $= 0$ then the volume of seawater in the segment is

$$V_{sn} = V_n \times \frac{S_n}{S_0} \tag{6.1}$$

and the volume of the freshwater in the segment is

$$V_{fn} = V_n - V_{sn} = V_n - V_n \times \frac{S_n}{S_0}$$

$$= V_n \frac{(S_0 - S_n)}{S_0}. \tag{6.2}$$

If the freshwater flow from the river is Q during a tidal cycle then the retention of the freshwater in the segment is

$$R_n = \frac{V_{fn}}{Q}$$

$$= \frac{V_n}{Q} \frac{(S_0 - S_n)}{S_0}$$

As the effluent is assumed to behave in the same way as the freshwater then the effluent retention will be the same as equation (6.3).

Considering the outfall segment N into which F is the volume of conservative effluent discharged during one tidal cycle. At equilibrium, the amount of effluent entering the segment must equal the amount being advected downstream during one tidal cycle.

Taking a mass balance across the discharge segment

$$F = \frac{C_N \times V_N}{R_N} \tag{6.4}$$

where C_N is the equilibrium concentration in segment N. Rearranging equation (6.4) gives

$$C_N = \frac{F \times R_N}{V_N}$$

$$= F \times \frac{V_N}{Q} \times \frac{(S_0 - S_N)}{S_0} \times \frac{1}{V_N}$$

$$= \frac{(S_0 - S_N)}{S_0} \times \frac{F}{Q} \tag{6.5}$$

For a conservative substance the amount conveyed downstream is always equal to the amount discharged at the outfall and the concentration in any downstream segment n is given by

$$C_n = \frac{(S_0 - S_n)}{S_0} \times \frac{F}{Q} \tag{6.6}$$

For upstream segments subject to dispersive processes only, the ratio of the upstream concentration C_m to the outfall concentration C_N will be the same as the ratio of the salinities of the two segments giving

$$\frac{C_m}{C_N} = \frac{S_m}{S_N}$$

and

$$C_m = \frac{S_m}{S_N} \times C_N \tag{6.7}$$

Using equations (6.6) and (6.7) together with knowledge of the uniform effluent flow, uniform freshwater flow, half-tide cross-sectional average salinity data, an estimate of the concentration profile of conservative substance can be made.

This simple approach has been extended to take account of non-conservative substances. However, it is not intended to consider this model in more detail but to look at the approach used for the Thames.

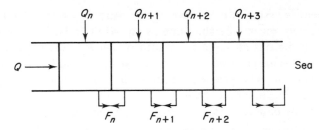

Fig. 6.6 A segmented-esturary model with interchange between segments

Again, complete mixing within segments is assumed with steady freshwater flow and effluent flow and an equilibrium salinity distribution. Segment lengths are chosen equal, to approximately one-fifth of a tidal excursion and the continuous effluent discharge for each tide is assumed to be distributed over a distance equal to the tidal excursion. Effluent is distributed in suitable proportions among those segments which normally receive additions during each tidal cycle.

Consider Fig. 6.6 in which the estuary has been divided into segments.

The effect of tidal mixing is represented by equal and opposite flows F between segments. Q, V, and S are the freshwater flow half-tide volume and half-tide salinity respectively.

By taking a mass balance for salt across each segment a series of equations are obtained in which the only unknowns are the values for F.

Segment n $\quad F_n(S_{n+1} - S_n) - (Q + Q_n)S_n = 0$

Segment $n + 1$ $F_{n+1}(S_{n+2} - S_{n+1}) + F_n(S_n - S_{n+1}) + (Q + Q_{n)}S_n$

$$- (Q + Q_n + Q_{n+1})S_{n+1} = 0$$

Segment $n + 2$ $F_{n+2}(S_{n+3} - S_{n+2}) + F_{n+1}(S_{n+1} - S_{n+2})$

$$+ (Q + Q_n + Q_{n+1})S_{n+1}$$

$$- (Q + Q_n + Q_{n+1} + Q_{n+2})S_{n+2} = 0 \tag{6.8}$$

Solving these equations simultaneously will give the values for F the unknown mixing coefficients. Having found the F values the model can be applied to any other dissolved or suspended substance.

If a substance C which decays exponentially is discharged into segment $n + 1$ then the rate of decay of the substance is $kC_{n+1}V_{n+1}$ where k is the decay rate constant. The mass balance equation for this segment is then

$$\frac{\partial}{\partial t}(C_{n+1}V_{n+1}) = F_n(C_n - C_{n+1}) + F_{n+1}(C_{n+2} - C_{n+1})$$

$$+ (Q + Q_n)C_n - (Q + Q_n + Q_{n+1})C_{n+1}$$

$$- kC_{n+1} V_{n+1} + L_a = 0 \tag{6.9}$$

where L_a is the mass added per tidal cycle. The solution to the mass balance equations for all segments yields the steady-state distribution for the concentration profile C.

This type of model would not be be expected to reproduce pollutant distributions with large concentration gradients because of the step length involved. Fortunately the distribution of effluents in many estuaries exhibit shallow gradients.

Refinements to the one-dimensional mixed segment model include increasing the number of layers to describe stratified flow and also the development of a time dependent model by advecting the boxes up and downstream in response to tidal movements.

6.3.4 Moving segment models

In this approach, the zone of interest is divided into a number of segments. As the tide ebbs and flows the segments move up and down the estuary changing their shape to maintain the volume within each segment constant. If the boundaries of the segments are moving with tidal velocity U_T then the volume of the segment will remain constant with time.

Also for any segment boundary X_i the volume upstream of that boundary

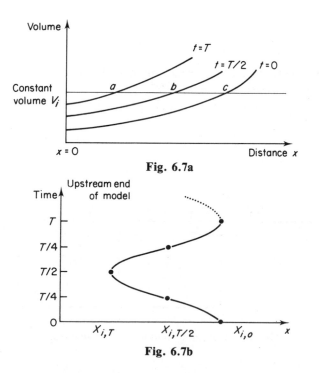

Fig. 6.7a

Fig. 6.7b

will remain constant with time. This fact can be used to calculate the boundary position at any point in time. A graphical representation is shown in Figure 6.7a and 6.7b.

Figure 6.7a is a plot of estuary volume against distance from the upstream boundary of the model for three separate points in time, $t = 0$, $T/2$ and T. Where $t = 0$ is the time of low tide, and $t = T/2$ and T are the times of high tide and the next low tide respectively. Let $X_{i,0}$ be the boundary of a particular segment at low tide and let V_i be the upstream volume of that boundary. From the previous discussion V_i is to remain constant for all t and may be plotted in Figure 6.7a as a horizontal line which intersects the volume plots at a, b and c. Perpendiculars dropped to the x-axis identify the position of the boundary at low tide, mean tide and high tide. If the volume/distance graph is plotted for intermediate times, a graph of time/distance can be constructed as shown in Figure 6.7b identifying the boundary at any time during the tidal cycle. This information can be computed for each segment boundary and hence the centre of each segment can be established at any point in time.

In Fig. 6.8 consider the boundary X_i between segment i and $i + 1$. The transport of material through the boundary is made up of two terms.

1. Due to freshwater flow, $Q_i \cdot C_i$ where $Q_i = A_i(U_i - U_{Ti})$

    $A_i = $ cross sectional area at i

    $U_i = $ velocity of freshwater

    $U_{Ti} = $ velocity of the tide

2. Due to longitudinal dispersion, simulated by assuming equal and opposite flow between segments, $F_i(C_i - C_{i+1})$.

It is now possible to write a mass balance for segment i giving

$$V_i \frac{dC_i}{dt} = Q_{i-1} C_{i-1} + F_{i-1}(C_{i-1} - C_i) - Q_i C_i$$

$$+ F_i(C_{i+1} - C_i) - K_i V_i C_i + L_{ai} \qquad (6.10)$$

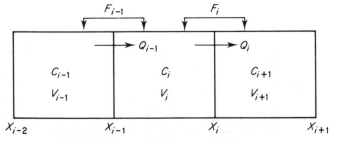

Fig. 6.8 Constant volume moving segments

where

$$L_{ai} \text{ is the discharge load}$$

$$K_i \text{ is the decay rate}$$

This approach has reduced the one-dimensional partial differential equation of the general form described in equation (5.16) to an ordinary differential equation (6.10) by changing the reference frame from a fixed spatial grid to a frame moving with the tidal motion. The model is dynamic in that changes of concentration with respect to time and distance can be simulated.

Essentially the computer program is split into two distinct parts. The first stage is to start with known conditions of segment size and position at a particular stage in the tidal cycle with an initial concentration profile. After a known time increment Δt, a new tide height can be calculated using empirical data on tidal behaviour. From this the new tidal volume can be calculated from which the movement of segment boundaries can be determined. Having found the new position of the segments the second stage of the computation is to solve equation (6.10) for each segment.

The Crank–Nicholson approximation may be used to numerically solve equation (6.10) for small increments of time Δt. The time derivative dC_i/dt may be represented by a forward difference approximation over the time interval j to $j + 1$ by

$$\frac{C_i^{j+1} - C_i^j}{\Delta t}$$

The right-hand side of equation (6.10) can be approximated by the average of values at j and $j + 1$ time steps. In other words, the difference scheme is centred in time about the point $j + \frac{1}{2}$.

Equation (6.10) now becomes:

$$(V_i^{j+1}C_i^{j+1} - V_i^jC_i^j) = \frac{1}{2}\Delta t(Q_{i-1}^{j+1}C_{i-1}^{j+1} - Q_i^{j+1}C_i^{j+1}$$

$$+ F_{i-1}^{j+1}(C_{i-1}^{j+1} - C_i^{j+1}) + F_i^{j+1}(C_{i+1}^{j+1} - C_i^{j+1}) + K_iV_i^{j+1}C_i^{j+1})$$

$$+ \frac{1}{2}\Delta t(Q_{i-1}^jC_{i-1}^j - Q_i^jC_i^j + F_{i-1}^j(C_{i-1}^j - C_i^j) + F_i^j(C_{i+1}^j - C_i^j)$$

$$+ K_iV_i^jC_i^j) \tag{6.11}$$

Each equation for each segment has three unknown values of C_{i-1}^{j+1}, C_i^{j+1}, C_{i+1}^{j+1} and can be rearranged in the form:

$$\alpha_iC_{i+1}^{j+1} + \beta_iC_{i+1}^{j+1} + \gamma_iC_{i+1}^{j+1} = \delta_i(C_{i-1}^j, C_i^j, C_{i-1}^j) \tag{6.12}$$

These can be solved simultaneously using techniques described in the next section and the values obtained used as initial values for the next time increment. The solution is therefore iterative. At each increment of time the positions of the segments are calculated first and then the new concentrations in each segment.

The exchange coefficients F_i may be found using salinity data and assuming that there is a balance between the salt transported between the segments by the freshwater velocities and the exchange flows, e.g.

$$F_i(S_{i+1} - S_i) = Q_i S_i$$

$$F_i = \frac{Q_i S_i}{(S_{i+1} - S_i)} \qquad (6.13)$$

On substituting equation (6.13) into equation (6.11) and using the known salinity profile for initial conditions, the model may be run for a series of time steps to check that the salinity in each box remains constant in time.

One advantage for this type of model is that conceptually it is easy to visualize the representation of the advection and dispersion mechanisms. A second significant advantage is that this formulation reduces the tendency for numerical dispersion to occur compared with the fixed grid finite difference approach.

Models of this type have been successfully used on the Thames and Humber Estuaries.

6.3.5 Finite difference estuary models

Several models of one and two dimensions have been formulated based on the numerical solution of the convective diffusion equation using finite difference techniques. The following example describes the solution to the one-dimensional equation using an implicit difference approximation.

The general form of equation (5.16) may be used to describe the estuary situation.

$$\frac{\partial AC}{\partial t} = \frac{-\partial AUC}{\partial x} + \frac{\partial}{\partial x}\left(EA\,\frac{\partial c}{\partial x}\right) - KAC + L_a \qquad (6.14)$$

The cross-sectional area A, effective longitudinal dispersion coefficient E and advective velocity U are functions of time t and distance x. Figure 6.9 shows the distance time plane covered by a mesh of sides $\delta x = L$ and $\delta t = K$. The implicit solution steps through time by increments of K and at each new point in time evaluates the values of C for all mesh points along the x direction.

For each point on the $j+1$ row the derivatives in equation (6.10) are estimated using the values of the six points shown in Fig. 6.9. Thus for n

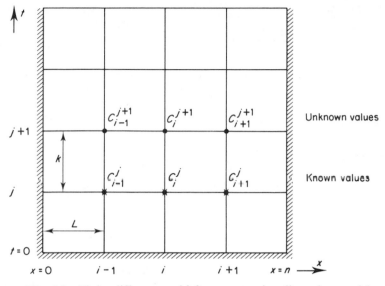

Fig. 6.9 Finite difference grid for an estuarine dispersion model

distance mesh points $n - 2$, simultaneous equations are developed each of which has three unknowns except for the equations at each end. They only have two unknowns at the $j + 1$ level because the boundaries of the mesh are chosen to produce known values of concentration at all time intervals. These known boundary values are essential for solving the simultaneous equations as shown below. The implicit solution is started by assuming an initial concentration distribution at $t = 0$. This particular formulation produces a dynamic solution which varies with time and the model is run until the particular dynamic effect of a transient load is represented or until a pseudo-steady state is reached for uniform loads at which concentrations do not vary between equivalent points in the tidal cycle.

A typical implicit approximation method (Crank – Nicholson) approximates derivatives in the x direction i.e. $\partial/\partial x$ and $\partial^2/\partial x^2$ by the mean of its finite difference representation on the $j + 1$ and the j rows.

In calculating C_i^{j+1} the time derivative $\partial/\partial t$ used to advance the computation is computed from a weighted average of the derivatives at the distance points $i - 1$, i and $i + 1$. The weighting used here is that giving a stable solution with minimal numerical dispersion which is insignificant if the value of E is reasonably large.

Each term in equation (6.14) is represented separately below by its difference approximation.

(a) *Time derivative*

$$\frac{\partial AC}{\partial t} \approx \frac{1}{6}\left(\frac{A_{i-1}^{j+1}C_{i-1}^{j+1} - A_{i-1}^{j}C_{i-1}^{j}}{k}\right)$$

$$+ \frac{2}{3}\left(\frac{A_i^{j+1}C_i^{j+1} - A_i^{j}C_i^{j}}{k}\right) + \frac{1}{6}\left(\frac{A_{i+1}^{j+1}C_{i+1}^{j+1} - A_{i-1}^{j}C_{i-1}^{j}}{k}\right) \qquad (6.15)$$

here the differences at $i-1$, i, $i+1$ are weighted $\frac{1}{6}$, $\frac{2}{3}$, $\frac{1}{6}$ respectively.

(b) *Advective term*

$$\frac{\partial UAC}{\partial x} \approx \frac{1}{2}\left(\frac{U_{i+1}^{j+1}A_{i+1}^{j+1}C_{i+1}^{j+1} - U_{i-1}^{j+1}A_{i-1}^{j+1}C_{i-1}^{j+1}}{2h}\right)$$

$$+ \frac{1}{2}\left(\frac{U_{i+1}^{j}A_{i+1}^{j}C_{i+1}^{j} - U_{i-1}^{j}A_{i-1}^{j}C_{i-1}^{j}}{2h}\right) \qquad (6.16)$$

equal weight is given to the j and $j+1$ differences which both use values at mesh points $i+1$ and $i-1$ divided by twice the x mesh distance to approximate the derivative at mesh distance i.

(c) *Dispersive term*

$$\frac{\partial}{\partial x} EA \frac{\partial C}{\partial x} \approx \frac{1}{h} \delta x \, (EA)_i \frac{1}{h} \delta x \, C_i^j \qquad (6.17)$$

where δx is called the central difference operator i.e.

$$\delta x \, C_i^j = C_{i+\frac{1}{2}}^j - C_{i-\frac{1}{2}}^j$$

and

$$\frac{\partial C}{\partial x} \approx \frac{\delta x C_{ij}}{h}$$

the derivative being evaluated at the point i, j.

Hence

$$\frac{1}{h} \delta x \left[(EA)_i \frac{1}{h} \delta x \, C_i \right] = \frac{1}{h} \delta x \left[(EA)_i \frac{1}{h}(C_{i+\frac{1}{2}} - C_{i-\frac{1}{2}}) \right]$$

$$= \frac{1}{h^2} \left\{ [(EA)_{i+\frac{1}{2}}C_{i+1} - (EA)_{i-\frac{1}{2}}C_i] \right.$$

$$\left. - [(EA)_{i+\frac{1}{2}}C_i - (EA)_{i-\frac{1}{2}}C_{i-1}] \right\} \qquad (6.18)$$

Thus the dispersive term is approximated using values of C at $i-1$, i, and $i+1$ but E and A are evaluated at the half mesh point in the x direction.

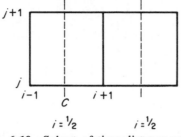

Fig. 6.10 Subset of time distance mesh

Now

$$(EA)_{i+\frac{1}{2}} = \frac{(EA)_{i+1} + (EA)_i}{2}$$

and

$$(EA)_{i-\frac{1}{2}} = \frac{(EA)_i + (EA)_{i-1}}{2}$$

Substituting in equation (6.18) gives

$$\frac{1}{h^2} \left\{ \left[\left(\frac{(EA)_{i+1} + (EA)_i}{2} \right) C_{i+1} - \left(\frac{(EA)_i + (EA)_{i-1}}{2} \right) C_i \right] \right.$$

$$\left. - \left[\left(\frac{(EA)_{i+1} + (EA)_i}{2} \right) C_i - \left(\frac{(EA)_i + (EA)_{i-1}}{2} \right) C_{i-1} \right] \right\} \qquad (6.19)$$

and grouping terms gives

$$\frac{1}{h^2} \left[\left(\frac{(EA)_{i+1} + (EA)_i}{2} \right)(C_{i+1} - C_i) - \left(\frac{(EA)_i + (EA)_{i-1}}{2}(C_i - C_{i-1}) \right) \right]$$

$$(6.20)$$

Now the dispersive derivative is approximated in equation (6.20) by the variable coefficients E and A evaluated at the mesh points.

As mentioned earlier, the Crank–Nicholson approximation uses equally weighted differences at the j and $j+1$ time intervals so that

$$\frac{\partial}{\partial x} EA \frac{\partial C}{\partial x} \approx \frac{1}{2} \cdot \frac{1}{2h^2} \left(\{ [(EA)_{i+1}^{j+1} + (EA)_i^{j+1}] C_{i+1}^{j+1} - C_i^{j+1}) \} \right.$$

$$- \{ [(EA)_i^{j+1} + (EA)_{i-1}^{j+1}](C_i^{j+1} - C_{i-1}^{j+1}) \} \right)$$

$$+ \frac{1}{2} \cdot \frac{1}{2h^2} \left(\{ [(EA)_{i+1}^j + (EA)_i^j (C_{i+1}^j - C_i^j) \} \right.$$

$$- \{ [(EA)_i^j + (EA)_{i-1}^j](C_i^j - C_{i-1}^j) \} \right) \qquad (6.21)$$

(d) *Reaction term*

$$kAC \simeq k \left[\frac{(AC)_i^{j+1} + (AC)_i^{j}}{2} \right]$$ (6.22)

This term uses the average of the C and A values at the j and $j+1$ row.

(e) *Load term*

$La = \Delta LA_i$ where ΔLa_i is the amount of effluent discharged over the time interval k at the distance meshpoint i.

The combination of difference terms produces a difference formula with three unkown values of C at the $j+1$ time step. Rearranging this expression by combining terms an equation may be produced with the three unknowns on the left-hand side and known values on the right-hand side, i.e.

$$- \alpha_i^{j+1} C_{i-1}^{j+1} + \beta_i^{j+1} C_i^{j+1} - \gamma_i^{j+1} C_{i-1}^{j+1} = \delta_i^j$$ (6.23)

where

> α and γ are functions of A, E, U at $j+1$
> β is a function of A and E at $j+1$
> δ is a function of A, E, U, C at j

Equation (6.19) is one of a system of n linear algebraic equations where $1 < i < n-1$ and C_0 and C_n are known from boundary conditions.

If

$$\alpha_i > 0 \qquad \beta_i > 0 \qquad \gamma_i > 0$$

and

$$\beta_i \geqslant (\alpha_i + \gamma_i)$$

then this system may be solved.

When written in matrix notation it may be seen that a tridiagonal form is produced. An efficient solution equivalent to Gaussian elimination is available to solve the system which may be written as

$$\begin{bmatrix} \beta_1 - \gamma_1 & & & & & \\ -\alpha_2 & \beta_2 & \gamma_2 & & & \\ & & \cdots & & & \\ & & & \cdots & & \\ & & & -\alpha_{n-2} & \beta_{n-2} & \gamma_{n-2} \\ & & & & -\alpha_{n-1} & \beta_{n-1} \end{bmatrix} \begin{bmatrix} C_1 \\ C_2 \\ . \\ . \\ C_{n-2} \\ C_{n-1} \end{bmatrix} = \begin{bmatrix} \delta_1 + C_0 \\ \delta_2 \\ . \\ . \\ \delta_{n-2} \\ \delta_{n-1} + \gamma_{n-1} \quad C_n \end{bmatrix}$$ (6.24)

Let us assume that $C_0 = 0$ and $C_n = K$.
The first equation may be written as

$$\beta_1 C_1 - \gamma_1 C_2 = \delta_1 \qquad (6.25)$$

and the second equation as

$$- \alpha_2 C_1 + \beta_2 C_2 - \gamma_2 C_3 = \delta_2 \qquad (6.26)$$

from equation (6.25)

$$C_1 = \frac{\delta_1 + \gamma_1 C_2}{\beta_1} = \frac{\gamma_1}{\beta_1} C_2 + \frac{\delta_1}{\beta_1} \qquad (6.27)$$

substituting for C_1 in (6.26) gives

$$- \alpha_2 \left(\frac{\gamma_1 C_2}{\beta_1} + \frac{\delta_1}{\beta_1} \right) + \beta_2 C_2 - \gamma_2 C_3 = \delta_2 \qquad (6.28)$$

rearranging for C_2

$$C_2 \left(\beta_2 - \frac{\alpha_2 \gamma_1}{\beta_1} \right) = \delta_2 + \frac{\alpha_2 \delta_1}{\beta_1} + \gamma_2 C_3$$

$$C_2 = \frac{\delta_2 + \alpha_2 (\delta_1 / \beta_1)}{\beta_2 - \alpha_2 (\gamma_1 / \beta_1)} + \frac{\gamma_2}{\beta_2 - \alpha_2 (\gamma_1 / \beta_1)} C_3 \qquad (6.29)$$

C_2 may now be substituted in the next equation and C_3 expressed in terms of C_4.

On inspection of equation (6.27) this may be rewritten as

$$C_1 = a_1 C_2 + b_1 \qquad (6.30)$$

where

$$a_1 = \frac{\gamma_1}{\beta_1} \quad \text{and} \quad b_1 = \frac{\delta_1}{\beta_1}$$

Now equation (6.29) may be rewritten as

$$C_2 = \frac{\delta_2 + \alpha_2 b_1}{\beta_2 - \alpha_2 a_1} + \frac{\gamma_2}{\beta_2 - \alpha_2 a_1} \cdot C_3 \qquad (6.31)$$

In general, an expression for C_i may be written in terms of C_{i+1} in the form

$$C_i = a_i C_{i+1} + b_i \qquad (6.32)$$

where

$$a_i = \frac{\gamma_i}{\beta_i - \alpha_i a_{i-1}}, \qquad b_i = \frac{\delta_i + \alpha_i b_{i-1}}{\beta_i - \alpha_i a_{i-1}}$$

With

$$C_0 = 0 \quad \text{then} \quad a_0 = 0 \quad \text{and} \quad b_0 = 0$$

hence

$$C_0 = a_0 C_i + b_0 \qquad (6.33)$$

equation (6.33) holds true for any value of C_i.

Starting with the first equation in the system, equation (6.32) can be used to successively eliminate C_i from the next equation in the system.

When the last equation is reached $C_n = K$ and

$$C_{n-1} = a_{n-1} K + b_{n-1} \qquad (6.34)$$

from which C_{n-1} can be calculated using the known values of a_{n-1}, K and b_{n-1}.

All the remaining C_i's can now be calculated by back substitution, e.g.

$$C_{n-2} = a_{n-2} C_{n-1} + b_{n-2}$$
$$\vdots$$
$$C_1 \quad = a_1 C_2 + b_1 \qquad (6.35)$$

To avoid substantial errors with this method it is necessary that

$$0 < a_i \leqslant 1 \quad \text{for} \quad 1 \leqslant i \leqslant n - 1$$

Once the values of C_i^{j+1} have been calculated, they become the initial values for the next time step and the process is repeated.

As mentioned earlier, α, β and γ are functions of the cross-sectional area A, dispersion coefficient E and velocity U. It is necessary that each of these is calculated at time $j + 1$ for each i mesh point in order that the system of equations described in equation (6.23), can be solved for the concentration distribution at $i + 1$.

The velocity U_i^{j+1} may be calculated in two ways. Firstly, U_i^{j+1} can be obtained from a solution of the equations of motion which may be developed in a similar way to the convective diffusion equation by balancing the forces acting on an element of volume of water. The equation of motion itself must be solved by a numerical technique similar to that described above. Using the velocity in the continuity equation, the tidal height can be calculated and hence the cross-sectional area at time level $j + 1$.

An alternative method of calculating U is to assume that the water level in the estuary is essentially horizontal and that it varies sinusoidally according to the relationship

$$h = h_{\text{mean}} - h_{\text{amp}} \cos \omega t \qquad (6.36)$$

where

    h_{mean} is the mean tide height
    h_{amp} is the tidal amplitude
    ω is $2\pi/12.4$ (assuming a tidal period of 12.4 hours)
    t is time taken as zero at low tide

For a given time interval k the new tide height at $j + 1$ can be calculated. Knowing the hydrographic data of the estuary it is possible to calculate the change in volume upstream of any mesh point i during the time interval k. This change in volume is equal to the flow through the cross-sectional area at i hence the velocity U_i^{j+1} can be calculated.

The one-dimensional dispersion coefficient for a homogeneous estuary zone may be calculated from an expression such as equation (6.37)

$$E = 63n \ U/R \qquad (6.37)$$

where

    n is Mannings Roughness Coefficient
    R is the hydraulic radius (m)
    E is in m^2/sec
    U is in m/sec

Equation (6.37) may be extended to zones which have longitudinal salinity gradients by including extra terms relating E to the gradient of the salinity profile.

The dispersion coefficient contains both advective and diffusive components because of the averaging process to reduce the convective diffusion equation from three to one dimension. It is, therefore, unique to the particular estuary under consideration. In practice, values included in the expression describing the dispersion coefficient are obtained from field observations using fluorescent or radio-active tracers. Even these may only give an order of magnitude result and a fine-tuning exercise is usually carried out by a curve-fitting process using the salinity profile. In other words, salinity data are collected from the estuary and the high- and low-tide salinity profiles are predicted over time using the finite difference scheme above with no decay or load terms. If the salinity profiles remain constant over a series of tidal cycles it may be assumed that the hydrodynamics of the estuary are correctly described. If the profiles alter significantly from the starting profile then adjustments are made to the dispersion coefficient.

Having successfully predicted the salinity distribution the model can be used to fit the DO sag curve and the UOD profile to existing data using the known existing discharge loads. The biochemical rate coefficients can be adjusted at this stage. When existing conditions can be satisfactorily represented predictions can be made of the UOD/DO profiles for different loading conditions.

6.4 SURVEYS AND DATA REQUIREMENTS

As mentioned above, it is not possible to produce a successful estuary model in isolation without collecting comprehensive information on estuary characteristics.

Initial surveys must be carried out to determine the appropriate type of model (1- or 2-D, etc.) and the appropriate boundary conditions.

Surveys are also required to produce basic data for the model:

1. Hydrographic data—length, cross-sectional area cumulative volume, tidal range.
2. Salinity data—estuary classification, determination of dispersion coefficient.
3. BOD—Profile.
4. DO—Profile.
5. Concentration of toxic materials.
6. Temperature profile.
7. Existing discharge loads.

The data are usually collected in the summer with low freshwater flow and high temperature which should theoretically represent the 'worst' conditions in the estuary. If water-quality objectives can be satisfied at this time of year they can usually be achieved at all other times.

A typical survey will last for about a week, usually covering a neap or spring tide period during which the tidal range does not vary significantly. Several survey stations are sited throughout the zone of interest and samples are taken at half hour intervals over a period of 13 hours which covers a complete tidal cycle. At each station samples are taken at several depths for each of the following parameters. Salinity, temperature, BOD, DO, Nitrates, velocity and toxics. From this data comprehensive 'snapshots' of the estuary can be built up.

This type of survey requires detailed organization and is expensive to mount.

6.5 BIBLIOGRAPHY

Ippen, T. (1966). *Estuary and Coastline Hydrodynamics.* McGraw-Hill. New York.
McDowell, D.M. and O'Connor, B.A. (1977). *Hydraulic Behaviour of Estuaries.* MacMillan, London.
Gameson, A.L.H. (1973). *Mathematical and Hydraulic Modelling of Estuarine Pollution.* Water Research Technical Paper No. 13 HMSO, London.

Chapter 7

Lake and Reservoir Modelling

A. JAMES

7.1 INTRODUCTION

The modelling of water quality in lakes and reservoirs is rather different from that in rivers and estuaries. The primary uses of the water—amenity, fisheries and abstraction are the same but the pollution pattern shows two major differences:

(a) Lakes and reservoirs rarely receive discharges of organic matter large enough to cause serious oxygen depletion.
(b) Due to the much greater retention time (and depth), lakes and reservoirs are dominated by planktonic organisms and are therefore more sensitive than rivers and estuaries to eutrophication.

The majority of models of lakes and reservoirs are therefore concerned with algal growth and inorganic nutrients. The concentrations of BOD and DO are of secondary importance as are the kinetics of bacteria (apart from nutrient recycle).

Other significant differences are the greater interest for lakes in the growth patterns of zooplankton both as algal predators and as food for fish. Also the differences in spatial heterogeneity of water quality—these tend to be horizontal and gradual in rivers whereas in lakes they are vertical and abrupt.

The quality of water in lakes and reservoirs is due to a combination of groups of factors namely:

(a) Influent quality and mixing pattern.
(b) Physical and chemical processes during storage.
(c) Biological growths and their role in the removal and release of substances.

Modelling water-quality changes in lakes therefore involves representation of all these factors, they are discussed separately below with the emphasis on algal growth and its relation with nitrogen and phosphorus. Most aspects of mixing, chemical kinetics and bacterial kinetics are described in Chapter 4.

7.2 FACTORS AFFECTING WATER QUALITY IN LAKES

7.2.1 Mixing patterns and influent quality

Retention time in reservoirs and lakes is of fundamental importance in determining mixing. Most reservoirs and lakes have a retention time of at least a week which is long enough to allow for complete horizontal mixing. They may, therefore, be represented as stirred tanks provided that the surface area is not very large. It is difficult to generalize due to local conditions but lakes with areas up to $100\,km^2$ are often well mixed, so that the situation may be represented as shown in Fig. 7.1.

The variation in water quality may be solved analytically for various simple situations e.g. for a step change in influent quality

$$C_E = C_I + (Co - C_I) \exp (-Q*t/V) \tag{7.1}$$

where

Co = starting concentration in lake
Q = flow rate
V = volume
t = time
C_I = concentration in influent
C_E = concentration in effluent
C_L = concentration in lake

But the analytical method has a very limited range of application and cannot handle situations where two or more factors are varying with time.

A more general approach is to draw up a mass balance on the substance concerned and to use numerical methods to solve the resulting equation. The model may be represented as shown in Fig. 7.2.

There is also the possibility of vertical variation in quality due to thermal stratification. This rarely occurs in shallow lakes (i.e. $<10\,m$) but in deep lakes (i.e. $>30\,m$) it is usually present. In intermediate depths the tendency to

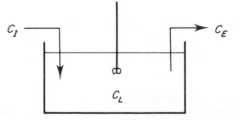

Fig. 7.1 Diagrammatic representation of a lake as a stirred tank reactor. (Note that $C_L = C_E$; $C_L \neq C_I$)

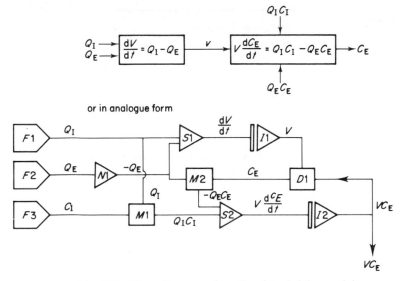

Fig. 7.2   Flow diagrams for stirred tank lake model

stratify depends upon the meteorological and topographical conditions. The stratification is seasonal in temperate lakes but is permanent in tropical lakes and reservoirs. Where permanent stratification occurs the density differences due to temperature are reinforced by salinity differences.

The modelling of stratification has not yet become a precise predictive tool but nevertheless useful guidance on the thermal pattern in a lake may be obtained from the solution of the one-dimensional heat transfer equation. Assuming the isotherms are horizontal the equation may be written as follows:

$$\frac{\partial T}{\partial t} = \frac{\partial}{\partial z}\left(A\,\frac{\partial T}{z}\right) + \frac{\partial}{\partial z}\left(K\,\frac{\partial T}{\partial z}\right) + Q \tag{7.2}$$

where

> T = temperature
> t = time
> z = depth

A and K are coefficients of molecular and turbulent diffusion
> Q is the heat source which may be given by

$$Q = \frac{\partial \phi}{\partial z}\Big/\varrho c$$

where

> ϕ = flux of solar radiation
> ϱ = density of water
> c = specific heat

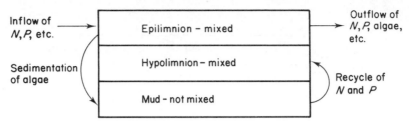

Fig. 7.3 Model of a stratified lake and sediment as a three-component system

The tendency to stratify is determined by the balance between the wind energy and the solar energy. Neglecting molecular diffusion the coefficient of eddy diffusion has been expressed as shown in the following equation

$$K = \frac{1}{P} \left(\frac{W^2}{\partial v/\partial z} \right) (1 + \sigma s)^{-n}$$

where the friction velocity (v) is related to the wind speed (w) and the stability parameter (S) may be expressed in the form of a Richardson number

$P = $ = Prandtl Number
σ and n are coefficients
     K *is Von Karman's constant*
     z_s is the height of the surface above datum ($z = 0$)

$$S = \left[-gK^2(z_s - z)^2/\varrho W^2 \right] \frac{\partial \varrho}{\partial z}$$

Where stratification occurs, the water body may be represented as a two-part system with the bottom deposits forming the third component with exchange only between epilimnion and mud; the mud to the hypolimnion. The inflow mixes only with epilimnion and this layer provides the outflow. Seiches cause the hypolimnion to mix and provided that there is enough wind energy to mix the epilimnion then the hypolimnion can be considered as fully mixed. Conditions in the mud are obviously not mixed but provided that the depth of mud is greater than 50 mm, then the rate of change within the mud is independent of depth and uniformity of composition may be assumed.

Unstratified conditions in lakes and reservoirs may be simply represented as a two compartment system.

The transition between stratified and unstratified periods in temperate lakes is fairly abrupt so that it may be represented as an instantaneous change.

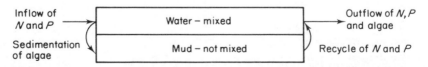

Fig. 7.4 Model of an unstratified lake and sediment as a two-component system

Exchange of heat or materials, other than as indicated above, between layers is so slow that it can be neglected in a model.

7.2.2 Physical and chemical processes during storage

The relatively long retention period of days, weeks or months in lakes and reservoirs gives ample time for water quality to change by both physical and chemical processes.

In the simplest lake models biological processes are not represented directly so that a process like nutrient uptake is simulated by a first-order removal. Chemical or pseudo-chemical processes may be represented in the usual way, e.g.

$$A + B \xrightarrow{K} C + D$$

the rate of reaction can be defined as

$$R = K_1(V)(C_A)(C_B) \tag{7.3}$$

so the mass balance becomes

$$V \frac{dC_{AE}}{dt} = Q_1 C_{A1} - Q_E C_{AE} - R$$

$$V \frac{dC_{BE}}{dt} = Q_1 C_{B1} - Q_E C_{BE} - R$$

$$V \frac{dC_{CE}}{dt} = R - Q_E C_{CE} \tag{7.4}$$

$$V \frac{dC_{DE}}{dt} = R - Q_E C_{DE}$$

which may be shown in analogue form as shown in Fig. 7.5.

The main physical process affecting water quality in lakes is sedimentation, although other processes like adsorption may be important for some pollutants.

Sedimentation—in falling freely through a quiescent liquid a particle accelerates to its terminal velocity V_s. If the particles are small and the settling velocities low the rate of settling obeys Stoke's law:

$$V_s = \left(\frac{s}{18}\right)\left(\frac{S_s - 1}{v}\right) d^2 \tag{7.5}$$

where

S_s = specific gravity of the solids
v = kinematic viscosity
d = characteristic diameter
q = the gravity constant

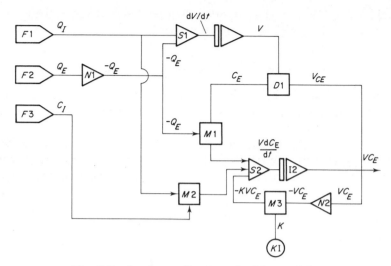

Fig. 7.5 Analogue diagram of a lake model

This can be incorporated into models and can be used to represent the sedimentation dead algal cells. But since retention times are long relative to the rate of sedimentation, a much simpler approach is to regard the sedimentation as instantaneous and the rate of removal as equal to the death rate.

7.2.3 Biological changes

Most of the important changes in water quality in lakes and reservoirs occur as a result of the activities of micro-organisms. It is therefore rarely possible to satisfactorily simulate those changes by purely chemical and physical models and it is important to find ways of modelling processes such as algal growth, nutrient uptake, nutrient recycle, predation, etc. These processes are discussed separately below followed by some suggestions for ways of incorporating these into lake models.

(a) *Algal growth*

This is a function of light, temperature and nutrients with losses due to death (and sedimentation), washout and predation. It is also important to distinguish between net and gross production:

$$\text{Net production} = \text{Gross production} - \text{respiration}$$
$$\text{Gross production} = \text{Photosynthesis}$$

The calculation of photosynthesis is the most complex aspect of lake models since it is so closely controlled by light intensity which is a function of the angle

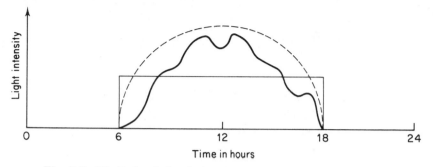

Fig. 7.6 Methods of simulating the daily pattern solar radiation

of incidence, altitude, latitude, cloudiness, depth, colour and turbidity and the length of day and night which varies seasonally. Some simplifying assumptions are required, especially in long-term models to avoid using an integration step shorter than one day. One way of doing this is to represent the daily light pattern as shown in Fig. 7.6.

The light pattern at the surface can be expressed in terms of day length and average intensity or as a semi-sinusoidal curve. The latter is more accurate because the rate of photosynthesis varies with light intensity but where detailed information about algal species is lacking (and consequently P_{max} and I_k) then the average daily intensity is likely to prove an acceptable approximation.

The basic equation relating photosynthesis to light intensity is generally given as follows:

$$P = P_{max} \frac{I}{I_k} \exp \left(1 - \frac{I}{I_k} \right) \tag{7.6}$$

$$P = \text{rate of photosynthesis}$$

where

P_{max} = maximum rate of photosynthesis
I = light intensity
I_k = optimum light intensity corresponding to P_{max}

For any incident intensity the radiation decreases with depth in a way which fits equation.

$$I(z + h) = I(z) \exp(-Keh)$$

where

$$Ke = \text{extinction coefficient}$$

$I(z)$ and $I(z + h)$ are light intensities at depth (z) and $(z + h)$ respectively.

It is possible to put all these factors together and obtain an expression from the daily photosynthesis in a water column integrated over depth. This is given by:

$$GP = \frac{N \cdot P_{max}}{Ke} \, 0.6 \left(1.33 \, \sinh^{-1}\phi \, - \frac{1}{\phi} \, (N1 + \phi^2) - 1 \right) \qquad (7.7)$$

where

GP = gross daily production per unit area
    = day length
N = algal concentration

$$\phi = \frac{I}{I_k}$$

The rate of photosynthesis increases with temperature up to an optimum and thereafter decreases. This may be expressed in a similar way to the light response as shown in equation (7.7)

$$Pt = P_{max} \, \frac{t}{t_{opt}} \, \exp \left(1 - \frac{t}{t_{opt}} \right) \qquad (7.8)$$

where

P_t = rate of photosynthesis at temperature t
t_{opt} = optimum temperature giving P_{max}

In many reservoirs and lakes the above equations are not adequate for predicting the growth of algal populations because of the limitations imposed by lack of nutrients. This additional feature can be incorporated by using a Michaelis–Menten equation for the limiting nutrients

$$P_L = P \left(\frac{N}{K_s + N} \right) \qquad (7.9)$$

Where

P_L = rate of growth with nutrients limitation
P = rate of growth in absence of nutrients limitation
N = concentration of limiting nutrients

but this approach fails to take account of the ability of algae to store nutrients and equation (7.8) therefore tends to predict prematurely the end of an algal bloom. A more satisfactory simulation can be obtained from a two-compartment model in which nutrients are taken up at a rate given in equation (7.9)

$$U = U_{max} \, \frac{(Q_m - Q)}{(Q_m - K_Q)} \left(\frac{N}{K_s + N} \right) \qquad (7.10)$$

where

U and U_{max} are the actual and maximum up take rates of nutrients
Q and Q_m are the quantity stored in all and the maximum storage

and the growth rate is determined by the stored nutrient according to equation (7.10)

$$G = G_{\max} \left(\frac{Q - K_Q}{Q} \right)$$ (7.11)

where

    G is the growth rate
    K_Q is a constant representing the minimum value of Q in the cell

(b) *Respiration*

In algal cells a significant proportion of the organic material produced by photosynthesis is used in respiration. Fortunately, the rate of respiration may be represented very simply as:

$$\frac{dN}{dt} = rN$$ (7.12)

where

    $r = $ *rate of respiration*
    $N = $ algal concentration

This equation needs a temperature correction where large variations in temperature occur. The correction is similar to that used for photosynthesis.

(c) *Sedimentation*

Some algal cells appear to be able to control their buoyancy but many, especially diatoms, have a tendency to sink and all species cease to be buoyant when dead. Sedimentation is therefore an important process in removing algae.

There is a complication in representing sedimentation because where the algae are still alive resuspension and re-growth is possible. The hydraulics are also complicated since shape factors for algae are not easily defined and are constantly changing due to small scale turbulence.

The simplest basis to represent sedimentation losses is a constant portion of the standing crop.

$$S = K_{\text{sed}} N$$ (7.13)

where

    $S = $ rate of sedimentation
    $K_{\text{sed}} = $ sedimentation coefficient

(d) Predation

In many reservoirs and lakes the effects of predation may be the major factor controlling the population levels of algae. Preferential grazing of some species may occur but in long-term models the composition of algae and predators is likely to change so it is best to make the representation very simple

$$\frac{dN}{dt} = -g\,NZ \tag{7.14}$$

where

g = filtering rate of predators
Z = predator concentration

The use of equation (7.13) obviously requires a prediction of the predator population. The kinetics of the predators can be based on the concept of yield and the rate of utilization of phytoplankton. This therefore becomes

$$\frac{dZ}{dt} = \frac{-g\,NZ}{Yz} \tag{7.15}$$

where
Yz = yield of predators per unit mass of algae consumed

An alternative approach that has been used is to regard the growth of zooplankton in terms of Michaelis–Menten kinetics with the algae as the substrate. The equation for growth therefore becomes:

$$\frac{dZ}{dt}\,j = \left[\sum_{i=1}^{m} \mu_{i,\,j}\,Y_{i,\,g}\,B_j\left(\frac{N_i}{B_j + N_i}\right)\right] g_j\,Z_j \tag{7.16}$$

where

μ_i = growth rate of species Z_j on alga N_i
i = yield coefficient
g = predation coefficient for species Z
B = Michaelis–Menten coefficient

The predation of higher trophic levels may be represented using similar equations but the complexities increase due to variation in fecundity and survival at different ages (see Chapter 4).

(e) Nutrient recycle

This occurs as a result of the bacterial decomposition of algal cells in the bottom mud. The rate of regeneration is therefore primarily dependent upon the concentration of organic matter in the mud. Since this is almost solely derived from the sedimented algae, the rate of regeneration can be simply represented as some function of the rate of sedimentation.

$$R = f(S) \qquad (7.17)$$

where

$$s = \text{rate of sedimentation}$$

When the limiting nutrient is phosphorus, the value of f is in two ranges— very low or very high depending whether the mud is aerobic or anaerobic. Under aerobic conditions most of the regenerated phosphorus (95–99%) is locked up in Ferric iron complexes. If the overlying water becomes anaerobic then the complexes break down and the value of f approximates to unity. It is therefore necessary to carry out an oxygen balance on the overlying water (see (f)).

(f) Oxygen balance

In the case where a lake or reservoir stratifies the oxygen balance is on the hypolimnion, this is effectively sealed from reaeration so the balance is merely

$$\text{DO}_t = \text{DO}_0 - \frac{R * K_D * SA}{V}$$

where

$\quad$ DO_0 and DO_t $\quad$ are the oxygen concentrations at the beginning and at any time t

$\quad$ K_D converts the rate of benthal bacterial activity into oxygen units

$\quad$ SA and V_H are the surface area of mud and the volume of the hypolimnion

Where the water body is not stratified, the oxygen balance is much more complex since other processes like photosynthesis, respiration, re-aeration and inflow and outflow are important. Fortunately, in these circumstances it is very rare for the lake water to become so deoxygenated as to cause liberation of phosphorus from Ferric complexes.

7.3 LAKE MODELS

Lake models may be classified into the following two categories:

(a) Long-term planning models whose aim is to forecast the increase in algal growth in a reservoir or to predict the effect of some management decision e.g. water transfer, watershed management, etc.

(b) Short-term operational models which are concerned with the day-to-day control of algal growth in an already eutrophic reservoir by induced circulation, algicides, etc.

Although the basic structure of these two types of model is similar, there are sufficient differences in detail to make it worthwhile discussing them separately.

7.3.1 Long-term planning models

The long-term changes in the productivity of a lake or reservoir may be represented diagrammatically as shown in Fig. 7.7.

Most models are concerned with the period of increasing productivity as the lake passes from oligotrophy to eutrophy and aim at forecasting the rate of change. The models vary in the hydraulics complexity and may be classified as

(a) Zero dimensional—whole water body regarded as a stirred tank reactor.
(b) Compartment model—water body divided into two horizontal layers and may also be divided into segments where lateral heterogeneity occurs.
(c) One-dimensional—Usually the dimension is vertical and allows for concentration gradients, light intensity gradients, etc.
(d) Two- and three-dimensional—usually cater for vertical variations with longitudinal and or lateral variations where these are important.

The choice depends upon the size of the impoundment, the precision required and the data available. It should be remembered that an increase in one dimension increases the data requirements by an order of magnitude.

The following example is a compartment model in which the water body is

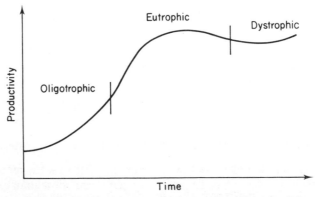

Fig. 7.7   Long-term changes in the productivity of a lake

divided into epilimnion and hypolimnion during the summer period and is fully mixed during the winter. The length of summer and winter are arbitrarily fixed at 165 days and 200 days respectively. The transition between stratified and unstratified conditions is regarded as instantaneous. Starting from the beginning of summer, the equations may be written in the form of daily balances on algae and nutrients in the epilimnion and dissolved oxygen and nutrients in the hypolimnion, as follows:

Epilimnion

$$\frac{dN}{dt} = GP - R - \text{sed} - \frac{Q_2 * N}{V_E} \qquad \text{algal balance} \qquad (7.18)$$

where

$$GP = \frac{N * P_{\max} * \Delta}{K_e} \, 0.6 \, [1.33 \, \sinh^{-1} \phi - \frac{1}{\phi} N(1 + \phi^2)]$$

$$- 1 \quad \text{algal production per day}$$

$$\phi = \frac{I}{I_k}$$

The light intensity is usually given as an average daily intensity for each month.

$$R = -r * N \qquad \text{respiration per day} \qquad (7.19)$$

$$\text{sed} = K_{\text{sed}} * N \qquad \text{sedimentation per day} \qquad (7.20)$$

$$Q_2 = \text{daily outflow and } V_e = \text{volume of epilimnion} \qquad (7.21)$$

$$\frac{dE_2}{dt} = \frac{Q_1}{V_E} * C_1 - \frac{Q_2}{V_E} * C_2 - u \qquad \text{nutrient balance} \qquad (7.22)$$

where

$$\frac{Q_1}{V_E} * C_1 = \text{daily in flux of nutrient}$$

$$\frac{Q_2}{V_E} * C_E = \text{daily out flux of nutrient}$$

and

$$u = u_{\max} \left(\frac{Q_m - Q}{Q_m - K_Q} \right) \left(\frac{C_2}{K_s + C_2} \right) \qquad \text{uptake rate per day}$$

where Q_m and Q are the maximum and current amounts of nutrient stored and K_Q is the Michaelis–Menten coefficient

Hypolimnion

$$\frac{dC_3}{dt} = \frac{f(\text{sed})}{V_4} \qquad (7.23)$$

where

C_3 = concentration of nutrients in hypolimnion
V_4 = volume of hypolimnion

$$\frac{dDO_H}{dt} = \frac{R * K_D * SA}{V_H} \qquad (7.24)$$

Each day during the summer period the mass balances are calculated and the concentrations of algal, nutrients and oxygen are adjusted.

At the end of the summer period the water body becomes fully mixed and the resulting concentrations of nutrients and algae become:

$$N_L = \frac{N_E * V_E + N_H * V_H}{V_E + V_H} \qquad (7.25)$$

$$C_L = \frac{C_E * V_E + C_H * V_H}{V_E + V_H} \qquad (7.26)$$

These are the starting concentrations for the winter period, during which the daily balances are given by:

$$\frac{dN_L}{dt} = GP - R - \text{sed} - \frac{Q}{V_L} * N \qquad (7.27)$$

$$\frac{dC_2}{dt} = \frac{Q_1}{V_L} * C_1 - \frac{Q_2}{V_L} * C_2 - u + f(\text{sed}) \qquad (7.28)$$

A dissolved oxygen balance is not normally required. At the end of the winter period the lake is divided arbitrarily into epilimnion and hypolimnion and the cycle begins again.

The flow diagram for this type of model is shown in Fig. 7.8 and a listing of the program is given in the Appendix to the chapter.

7.3.2 Short-term operational models

The two major differences between short-term and long-term models are as follows:

(a) In short-term models it is possible to represent the kinetics of the dominant species and to give similar detailed treatment to other inputs like solar radiation, temperature, nutrient inflow, etc.

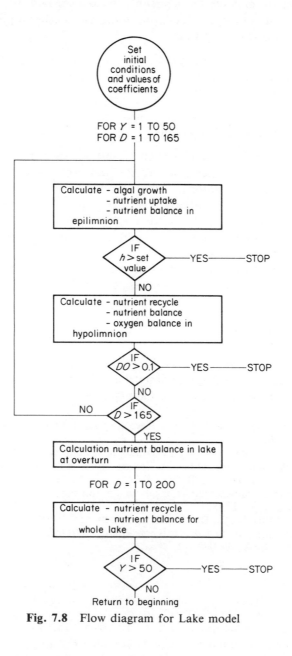

Fig. 7.8 Flow diagram for Lake model

(b) In short-term models it is possible to take account of predation and to take account of succession in the dominant species of prey and predator which will vary with time.

Because short-term models produce forecasts for weeks ahead rather than years, it is possible to reduce the size of the integration step below a day, which reduces the numerical errors.

7.3.3 Tropical lakes and reservoirs

There is a tendency for deep lakes and reservoirs in tropical areas to stratify permanently. In these circumstances the water quality in the hypolimnion becomes significantly different from the epilimnion such that chemical stratification is superimposed upon thermal stratification. The hypolimnion is permanently anaerobic so that nutrient release from dead algae is completely recycled. Return of nutrients to the epilimnion is however slow since it relies upon molecular diffusion.

Modelling of permanently stratified lakes may be carried out by suitable amendment to Fig. 7.8 as shown in Appendix 2.

7.4 BIBLIOGRAPHY

Unfortunately the literature on lake modelling is relatively sparse. The following publications will be helpful in model development and guidance on kinetic coefficients can be obtained from Lehman.

Scavia, D. and Robertson, A. (1979). *Lake ecosystem modelling*. Ann Arbor Science, Ann Arbor, Mich.

Steel, J. A. (1978). 'Reservoir Algal Productivity', Chapter 5. In *Mathematical Models in Water Pollution Control* (ed. A. James). John Wiley & Sons, Chichester.

Lehman, J. T., Botkin, D. B. and Likens, G. E. (1975). The assumptions and Rationales pf a computer model of phytoplankton population dynamics, *Limnology and Oceanography* **20** (3).

Zison, S. W., Mills, W. B., Deimer, D. and Chen, C. W. (1978). *Rates, Constants and Kinetic Formulations in Surface Water Quality Modelling*. Environmental Protection Agency. EPA-600/3-78-105.

Notes

K = Extinction coefficient (m^{-1})

$K1$ = Michaelis–Menten coefficient for phosphorus (mg l^{-1})

$K2$ = Michaelis–Menten coefficient for nitrogen (mg l^{-1})

$I1$ = Light intensity corresponding to maximum rate of photosynthesis (gm cal per cm^2 per hour)

P = Maximum rate of photosynthesis (μg of chlorophyll a per μg of chlorophyll a per litre per hour)

U = Proportion of sedimented nutrients that are recycled (dimensionless)

N = Algal concentrations (μg chlorophyll a per litre)

V = Volume of lake (m^3)

CO = Concentration of phosphate phosphorus in lake (mg l^{-1})

LO = Concentration of inorganic nitrogen in lake (mg l^{-1})

A = Area of lake (m^2)

$V1$ = Volume of epilimnion (m^3)

$V2$ = Volume of hypolimnion (m^3)

Y = Years

$C(Y)$ = Concentration of phosphate phosphorus in influent (mg l^{-1})

$L(Y)$ = Concentration of inorganic nitrogen in influent (mg l^{-1})

IO = Light intensity (gm calories per cm^2 per hour)

T = Temperature ($^\circ$C)

Q = Flow into lake (m^3 per day)

G = Daily photosynthetic production (μg per litre of chlorophyll a)

R = Daily losses by respiration (μg per litre of chlorophyll a)

S = Daily losses by sedimentation (μg per litre of chlorophyll a)

$C3$ = Concentration of phosphate phosphorus in the hypolimnion (mg l^{-1})

$L3$ = Concentration of inorganic nitrogen in the hypolimnion (mg l^{-1})

$C2$ = Rate of change of concentration of phosphate phosphorus per day in the epilimnion (mg per litre per day)

$L2$ = Rate of change of concentration of inorganic nitrogen per day in the epilimnion (mg per litre per day)

The lake is assumed to be stratified for seven months (Line 100) during which period algal growth, nutrient uptake and nutrient exchange with inflow/outflow take place in the epilimnion only. Recycle can occur into the hypolimnion depending upon the value of U.

At the end of the stratified period, the lake mixes (Lines 390 and 395) and during the remaining five months growth, uptake, exchange and recycle affect the whole volume.

APPENDIX 1

Lake model with temporary stratification

```
1    K = .4
2    K1 = .01
3    K2 = .1
7    I1 = 200
8    P = 0.03
```

```
10    DIMI(12),T(12),Q(12)
25    PRINT"INPUT U"
26    INPUT U
30    PRINT"INPUTN"
35    INPUTN
37    V = 6000000000
40    PRINT"INPUT CO"
41    INPUT CO
42    PRINT"INPUT LO"
43    INPUT LO
59    A = 150000000
70    V1 = 0.4*V
72    V2 = 0.6*V
75    FOR Y = 1 TO 10
80    READ  C(Y),L(Y)
85    DATA 1,5,1,5,1,5
86    DATA 1,5,1,5,.05,2
87    DATA .05,2,.05,2
88    DATA .05,2,.05,2
90    NEXT Y
92    FOR M = 1 TO 12
93    READ I(M),T(M),0(M)
94    DATA 300,5,50E05
95    DATA 300,5,30E05
96    DATA 330,6,30E05
97    DATA 350,6,20E05
98    DATA 350,6,20E05
99    DATA 330,6,25E05
100   DATA 300,5,34E05
101   DATA 280,4,50E05
102   DATA 270,3,100E05
103   DATA 270,3,90E05
104   DATA 280,4,150E05
105   DATA 300,5,92E05
108   NEXT M
109   PRINT"YEAR = ";Y
110   FOR Y =  1 TO 10
111   PRINT"YEAR = ";Y
112   FOR M = 1 TO 7
120   FOR D = 1 TO 30
150   C3 = CO
160   LE = LO
210   IF CO < 0.008THEN 300
```

```
212 IF LO < 0.05  THEN 300
220 G = (N*P*T(M)/K)*(EXP(1 – I(M)/I1))*(CO/(CO*K1))*(LO + K2))
230 R = 0.1*P*N*24
240 S = 0.01*N
245 C3 = C3 + ((S*U*0.003*V1)/V2)
246 L3 = L3 + ((S*U*0.01*V1)/V2)
250 GOTO 310
300 G = 0
310 C2 = (Q/V1)*(C(Y) – CO) – 0.0001*N
315 L2 = (Q(M)/V1)*(L(Y) – LO) – 0.0004*N
320 N = N + G – R – S
330 CO = CO + C2
335 LO = LO + L2
337 IFCO < OTHEN CO = 0
338 IFLO < OTHEN LO = O
339 IF N = < 0.1THEN N = 0.1
350 NEXT D
360 PRINT"N = ","PHOS = ";CO
365 PRINT"NO3 = ",LO
380 NEXT M
390 CO = (V1*CO + V2*C3)/V
395 LO = (V1*LO*V2*L3)/V
400 FOR M = 1 TO5
410 FOR D = 1 TO 30
420 G = (N*P*T(M)/K)*(EXP(1 – I(M)/I1))*(CO/(CO*K1))*(L0/(L0 + K2))
430 R = 0.1*P*N*24
440 S = 0.01*N
450 C2 = (Q(M)/V)*(C(Y) – CO) – 0.0001*N + S*U*.001
460 L2 = (Q(M)/V)*(L(Y) – L0) – 0.0004*N + 6*S*U*.001
470 N = N + G – R – S
480 CO = CO + C2
490 L0 = L0*L2
500 IFCO < 0THENCO = 0
510 IFLO < 0THENL0 = 0
520 IFN < 0.1THEN N = 0.1
530 NEXT D
540 PRINT"N = ",N
542 PRINT"PHOS = ",CO
544 PRINT"NO3 = ",L0
550 NEXT M
680 NEXT Y
690 DATA 300,5,50E05
700 DATA 300,5,30E05
```

```
710 DATA 330,6,30E05
720 DATA 350,6,20E05
730 DATA 350,6,20E05
740 DATA 330,6,25E05
750 DATA 300,5,34E05
760 DATA 280,4,50E05
770 DATA 270,3,100E05
780 DATA 270,3,90E05
790 DATA 280,4,150E05
800 DATA 300,5,92E05
810 END
```

APPENDIX 2

Lake model with permanent stratification

```
10   REM LAKE MODEL
20   REM INPUT CONSTANTS
30   PRINT"INPUT P,I1,K,K1,V,Q,P2,CO,S,N"
35   INPUT P,I1,K,K1,V,Q.P2,CO,S,N
40   FOR Y = 1 TO 20
50   PRINT"YEAR = ":Y
60   REM SUMMER CALCULATIONS
70   V1 = 0.4*V
80   V2 = 0.6*V
90   C3 = CO
100  FOR D = 1 TO 15
110  FOR W = 1 TO 10
120  P1 = 0.005*P2
130  C3 = C3 + P1/V2
140  P2 = P2 - P1 + S/100
150  NEXT W
160  NEXT D
170  REM EPILIMNION CALCULATIONS
180  FOR D = 1 TO 5
190  READ I0.T
200  FOR M = 1 TO 30
210  IF CO < 0.05 THEN 300
220  G = (N*P*T/K)*(EXP(1 - I0/I1))*(CO/(CO + K1))
230  R = 0.1*P*N*24
240  S = 0.01*N
250  GOTO 310
300  G = 0
```

```
310 C2 = Q/V1*C1 – Q/V1*CO – 0.001*N
320 N = N + G – R – S
325 IFN < 0THENN = 0
330 CO = CO + C2
340 NEXTM
350 PRINT"N = ";N,"PHOS = ";CO
360 NEXT D
370 REM WINTER CALCULATIONS
380 CO = (CO*V1 + C3*V2)/V
390 FOR 0 = 1 TO 200
400 P1 = P2*0.005
410 P2 = P2 – P1
420 C4 = Q/V*C1 – Q/V*CO + P1/V
430 CO = CO + C4
440 NEXT D
450 PRINT"PHOS CONC AT YEAR END = ":CO
460 N = 10
470 RESTORE
480 NEXT Y
490 DATA 30,5,40,6,50,7,60,8,50,7
500 END
```

For explanation of symbols see Notes in Appendix 1

Chapter 8

Mathematical Models of the Discharge of Wastewater into a Marine Environment

R. E. FEATHERSTONE

8.1 INTRODUCTION

A common method of wastewater disposal is to release it at some depth below the surface into a large body of water such as a lake or the sea by pipeline. The wastewater is released from the outfall either from a single open end or through a number of exit ports spaced along the pipeline in order to spread the release over a larger area.

The diameter of the outfall pipe is determined by the rate of flow with a constraint of achieving a self-cleansing velocity under dry-weather flow conditions. A figure of $0.76\,\mathrm{msec}^{-1}$ may be appropriate to raw sewage and $0.5\,\mathrm{msec}^{-1}$ in cases where the sewage is discharged after primary settlement. The exit velocity, and direction, of the discharge from a single or multiple port outlet, i.e. the 'jet velocity' affects to some extent the spread, and hence dilution, of the wastewater in the ambient liquid. Recent field investigations into the unsatisfactory performance of some outfalls have revealed the establishment and growth of marine mollusc colonies inside the outfalls and outlet

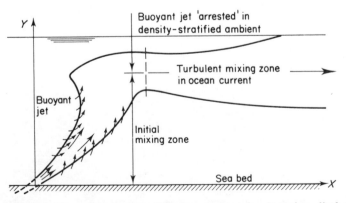

Fig. 8.1 Buoyant jet and turbulent diffusion zones of submarine discharge from outfall

ports probably due to the flow of sea water into the outfall. This could be caused by low jet velocities in relation to sea currents and wave action.

When wastewater is discharged from a pipe or diffuser into sea water the jet moves upwards due to the buoyant force which is proportional to the difference in density between the wastewater and the surrounding sea water (Fig. 8.1)

Dilution occurs as the 'buoyant' jet rises due to 'entrainment' of the sea water at the edge of the jet. The density differential therefore decreases with height and the jet may either become 'entrapped' below the surface or reach the surface depending on the depth of water above the outfall, the density stratification and velocity of the ambient and the hydraulic condition of the exit jet.

The dilution of the buoyant jet is referred to as 'initial dilution' or 'initial mixing'. Dilutions of up to 100 : 1 can be achieved in the initial mixing zone and the aim of outfall design is to achieve as high a figure as possible in order to avoid the formation of a visible slick. Also, since the further dilution of the buoyant plume in ocean currents is much smaller and cannot be controlled, momentum and buoyancy cease to have any major effect and the dispersion process is governed by environmental turbulence, ambient currents, and air–sea interactions.

8.2 INITIAL DILUTION IN THE BUOYANT JET

Studies of buoyant jets have extended over the past five decades beginning with the development of empirical relationships by Rawn and Palmer for the dilution of a sewage field from a point source on the sea bed. Morton introduced the entrainment assumption and further developments were carried out by Fan and Abraham. The models described here are based on the method of analysis of Fan in which the variables contained in the flow equations were further separated to describe the case of an arbitrarily stratified liquid (McBride).

The initial mixing region can be divided into two zones, the zone of flow establishment and the zone of established flow. In the zone of flow establishment the liquid is assumed to leave the orifice with a uniform velocity distribution across the jet. At the boundary, a cylindrical shear layer forms and develops into a turbulent mixing layer. The turbulent region spreads into the centre of the jet at approximately six jet diameters from the port. Velocity flux is assumed to be constant over this zone, i.e. no entrainment of ambient occurs. Immediately following the zone of flow establishment is the zone of established flow, in which the flow is termed a buoyant jet. In this zone the jet entrains ambient fluid continuously and thus increases in width and dilution. The density differential of the ambient liquid to the jet therefore decreases and in an ambient liquid having a lower density near the surface, the jet may become neutrally buoyant. Due to the momentum of the jet it will be

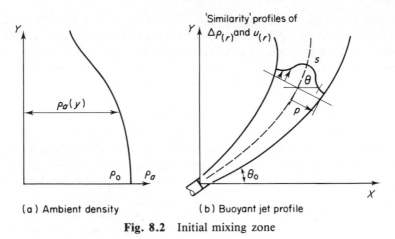

(a) Ambient density   (b) Buoyant jet profile

Fig. 8.2   Initial mixing zone

carried beyond this point becoming negatively buoyant eventually coming to rest and cascading down around the rising jet until reaching an equilibrium.

Consider the general case where the discharge is made at some angle into an arbitrarily stratified environment (Fig. 8.2).

Defining volume flux Q, momentum flux M and density deficiency flux B by:

$$Q = \int_A u(r)\ dA \tag{8.1}$$

$$N = \int_A \varrho u(r)^2$$

and since $\Delta\varrho \ll \varrho_0$

$$M = \int_A \varrho_0 u(r)^2\ dA = \varrho_0 \int_A u(r)^2\ dA \tag{8.2}$$

$$B = \int_A \Delta\varrho\ u(r)\ dA \tag{8.3}$$

Thus the total buoyance force is

$$f = \int g\ \Delta\varrho\ dA \tag{8.4}$$

From conservation,

$$\text{Mass:}\ \frac{dQ}{ds} = E \tag{8.5}$$

where

$$E = \text{rate of entrainment} \tag{8.5}$$

$$x\text{-Momentum: } \frac{d}{ds}(M\cos\theta) = 0 \tag{8.6}$$

$$Y\text{-Momentum: } \frac{d}{ds}(M\sin\theta) = f \tag{8.7}$$

Density deficiency:

$$\frac{d}{ds}\left(\int_A u(r)(\varrho_0 - \varrho)\,dA\right) = E(\varrho_0 - \varrho_a) \tag{8.8}$$

which can be written as

$$\frac{d}{ds}\left[\int_A u(r)(\varrho_0 - \varrho_a)\,dA + \int_A u(r)\,\Delta\varrho\,dA\right] = E(\varrho_0 - \varrho_a)$$

i.e.

$$(\varrho_0 - \varrho_a)\frac{dQ}{ds} + Q\left(-\frac{d\varrho_a}{ds}\right) + \frac{dB}{ds} = E(\varrho_0 - \varrho_a)$$

whence, using equation (8.5)

$$\frac{dB}{ds} = \frac{d\varrho_a}{ds}Q \tag{8.9}$$

Equations (8.5), (8.6), (8.7) and (8.9) are four equations as functions of s for the four unknowns Q, M, θ and B.

The local velocity $u(s, r)$ can be related to the jet velocity at the axis using the similarity profile and expressed in the form, $u(s, r) = u(s)\, f(\eta)$, where $\eta = r/b(s)$ in which b is a characteristic jet width.

Similarly, the density deficiency $\Delta\varrho$ can be related to that at the axis by

$$g\,\Delta\varrho(s, r) = g\,\Delta\varrho(s)\, f(\xi) \quad \text{where } \xi = \frac{r}{\lambda b}$$

in which λ is a coefficient which accounts for the greater spread of buoyancy than mass.

Experiments have demonstrated Gaussian similarity profiles for example

$$f(\xi) = e^{-\xi^2}$$

whence

$$u(s, r) = u(s)\, e^{-r^2/b^2}$$

$$\Delta\varrho(s, r) = \Delta\varrho(s, 0)\, e^{-r^2/\lambda^2 b^2}$$

Therefore

$$Q = \int_{r-0}^{b} u(s, 0)\, e^{-r^2/b^2}\,dA = \pi u b^2$$

where

$$u = u(s, 0)$$

Similarly, writing $\Delta\varrho = \Delta\varrho(s, 0)$.

Using the entrainment assumption,

$$E = P_r \alpha u$$

where P_r is the jet perimeter and α the entrainment coefficient.

$$B = \frac{\lambda^2}{1 + \lambda^2} \pi u b^2 \Delta\varrho$$

$$M = \frac{\pi}{2} u^2 b^2 \varrho_0$$

$$f = \pi \lambda^2 b^2 g \Delta\varrho$$

$$E = 2\pi\alpha u b$$

Substituting into equations (8.5) to (8.9) yields

$$\frac{du}{ds} = \frac{2g\lambda^2}{u} \frac{\Delta\varrho}{\varrho_0} \sin\theta \; - \frac{2u\alpha}{b}$$

$$\frac{d}{ds}(bu^2 \cos\theta) = 0$$

$$\frac{d}{ds}(bu^2 \sin\theta) = \sqrt{2g}\lambda b \frac{\Delta\varrho}{\varrho_0}$$

$$\frac{d}{ds}(bu \Delta\varrho) = bu \frac{\sqrt{1 + \lambda^2}}{\lambda} \frac{d\varrho_a}{dy} \sin\theta$$

$$\frac{dx}{ds} = \cos\theta; \qquad \frac{dy}{ds} = \sin\theta$$

Further manipulation of these equations yields:

$$\frac{du}{ds} = \frac{2g\lambda^2}{u} \frac{\Delta\varrho}{\varrho_0} \sin\theta - \frac{2u\alpha}{b} \tag{8.10}$$

$$\frac{db}{ds} = 2\alpha - \frac{b}{u^2} g\lambda^2 \frac{\Delta\varrho}{\varrho_0} \sin\theta \tag{8.11}$$

$$\frac{d\theta}{ds} = \frac{2g\lambda^2}{u^2} \frac{\Delta\varrho}{\varrho_0} \cos\theta \tag{8.12}$$

$$d\frac{\Delta\varrho}{ds} = \frac{1+\lambda^2}{\lambda^2} \sin\theta \, d \, \frac{\varrho a}{dy} - \frac{2\alpha\Delta\varrho}{\varrho} \tag{8.13}$$

$$\frac{dx}{ds} = \cos\theta \tag{8.14}$$

$$\frac{dy}{ds} = \sin\theta \tag{8.15}$$

Also

cub = constant where c = pollutant concentration

Thus

$cub = c_0 u_0 b_0$ where c_0, u_0 and b_0 are the initial conditions

which obtain at the end of the zone of flow establishment.

Computations begin at the downstream end of the zone of flow establishment. From experiments, Albertson found the length of this zone to be 6.2D.

From conservation of momentum

$$b_0 = D/\sqrt{2}$$

and for continuity of pollutant concentration

$$c_0 = \frac{1+\lambda^2}{2\lambda^2},$$

taking the outlet concentration as 1.0, and

$$\Delta\varrho_0 = \frac{1+\lambda^2}{2\lambda^2} (\Delta\varrho)_0,$$

where $(\Delta\varrho)_0$, is the density deficiency of the effluent.

Rouse found $\alpha = 0.082$ and $\lambda = 1.16$.

The solution of the six simultaneous differential equations (8.10) to (8.15) proceeds in small increments Δs along the jet, using a fourth order Runge–Kutta integration method. Data input consists of the initial jet diameter, velocity, density and inclination to the horizontal. The interval Δs and the ambient density at intervals Δy above the outlet are specified. Output quantities are x, y, u, θ. Width (defined as $2\sqrt{2}b$), and dilution (1/c), where $c = (c_0 u_0 b_0)/ub$. Computations cease when the surface is reached or when the inclination of the jet changes sign (entrapped jet).

The Runge–Kutta procedure is an explicit method yielding the values of the dependent variables, u, b, θ, Δp, x, y at the end of each finite increment in the independent variable s. Since, in this case Δx and Δy are determined directly from θ and Δs, the number of simultaneous ordinary differential equations is reduced to four.

Defining

$$\frac{du}{ds} = f_1(u, b, \theta, \Delta\varrho)$$

$$\frac{db}{ds} = f_2(u, b, \theta, \Delta\varrho)$$

$$\frac{d\theta}{ds} = f_3(u, b, \theta, \Delta\varrho)$$

$$\frac{d\Delta\varrho}{ds} = f_4(u, b, \theta, \Delta\varrho)$$

For example:

$$u(s + \Delta s) = u_{(s)} + \tfrac{1}{6}(k_{11} + 2k_{12} + 2k_{13} + k_{14})$$

where

$$k_{11} = \Delta s \, f_1(u_i, \ b_i, \ \theta_i, \ \Delta\varrho_i)$$

$$k_{12} = \Delta s \, f_1\left(u_i + \frac{k_{11}}{2}, \ b_i + \frac{k_{21}}{2}, \ \theta_i + \frac{k_{31}}{2}, \ \Delta\varrho_i + \frac{k_{41}}{2}\right)$$

$$k_{13} = \Delta s \, f_1\left(u_i + \frac{k_{12}}{2}, \ b_i + \frac{k_{22}}{2}, \ \theta_i + \frac{k_{32}}{2}, \ \Delta\varrho_i + \frac{k_{42}}{2}\right)$$

$$k_{14} = \Delta s \, f_1(u_i + k_{13}, \ b_i + k_{23}, \ \theta_i + k_{33}, \ \Delta\varrho + k_{43})$$

in which the subscript i indicates values at s. k_{21}, k_{31}, $k_{41} \ldots$ etc. are simultaneously and similarly obtained from the functions f_2, f_3 and f_4.

Similarly $b(s + \Delta s)$, $\theta(s + \Delta s)$ and $\Delta\varrho(s + \Delta s)$ are obtained and these values become u_i, b_i, θ_i and $\Delta\varrho_i$ at the next step.

The interval Δs should be small to maintain stability since the Runge–Kutta method is based on a truncated Taylor series.

A version of the program written in BASIC for use on a CBM Commodore PET is included as Appendix to this chapter. When using a microprocessor the independent variable should be very small due to the lower precision obtained in computing functions of the independent variables in the Runge–Kutta scheme. It was found that a value of $\Delta s = 0.001$ was necessary and the computing time is consequently lengthy. A mainframe or microcomputer is therefore recommended for such numerical techniques.

8.3 DIFFUSION OF NEUTRALLY BUOYANT JET IN AN OCEAN CURRENT

If the buoyant jet still possesses some buoyancy on reaching the surface it will spread on the surface and will be subject to the transporting effects of wind and waves. However, from experimental evidence obtained by Abraham and Brolsma, it appears that the buoyant spread may be practically ignored if the dilution achieved at this level is greater than $50:1$ since most discharge schemes should achieve this dilution in the initial mixing zone. Dilution in the essentially horizontal flow in ocean currents, following the end of the buoyant jet zone, will therefore be considered. In the practical case, measurements of sub-surface and surface currents should be carried out for each outfall scheme.

The mathematical model presented follows that of Koh, see McBride[3]. Assuming that the vertical and transverse velocity components are negligible in comparison with the steady-stream current, $Ua(x, y)$ and that longitudinal dispersion is also negligible the equation for convective diffusion is

$$Ua \frac{\partial c}{\partial x} = \frac{\partial}{\partial y}\left(K_y \frac{\partial c}{\partial y}\right) + \frac{\partial}{\partial z}\left(K_z \frac{\partial c}{\partial z}\right) - K_d c \tag{8.16}$$

in which c is the concentration of the pollutant and K_d is the decay coefficient. K_y and K_z are the dispersion coefficients.

The boundary conditions are

$$\frac{\partial c}{\partial y} = 0 \quad \text{at } y = 0 \text{ (bed)} \quad \text{and at } y = h$$

assuming zero exchange with the atmosphere, otherwise

$$K_y \frac{\partial c}{\partial y} = K_e c \quad \text{at } y = h.$$

Source conditions are determined from the buoyant jet calculations. Koh defined the source at $(0, y_0)$ with thickness h_0 and width $L = 4b_0$. The source may be at the surface ($y_0 = h$), or submerged. The source distribution is considered to be Gaussian in the z-direction and elliptical in the y-direction giving

$$c(0, y, z) = c(0, y_0, 0) \exp\left[-\frac{z^2}{b_0^2\left[1 - \left(\frac{2(y - y_0)}{h_0}\right)^2\right]^{1/2}}\right] \tag{8.17}$$

The number of independent variables in equation (8.16) is reduced from three to two by defining

$$c_0(x, y) = \int_{-\infty}^{\infty} c(x, y, z) \, dz \tag{8.18}$$

$$c_2(x, y) = \int_{-\infty}^{\infty} z^2 c(x, y, z) \, dz \tag{8.19}$$

assuming symmetry in the z direction.

Multiplying equation (8.16) by z^0 and z^2 and integrating over z,

$$U_a \frac{\partial c_0}{\partial x} = \frac{\partial}{\partial y}\left(K_y \frac{\partial c_0}{\partial y}\right) - K_d c_0 \tag{8.20}$$

$$U_a \frac{\partial c_2}{\partial x} = \frac{\partial}{\partial y}\left(K_y \frac{\partial c_2}{\partial y}\right) - K_d c_0 + 2K_z c_0 \tag{8.21}$$

with the boundary conditions

$$\frac{\partial c_0}{\partial y} = 0 \quad \text{and} \quad \frac{\partial c_2}{\partial y} = 0 \quad \text{at } y = 0 \quad \text{and} \quad y = h.$$

The source conditions become (equation (8.17)),

$$c(0, y) = c_0(0, y_0)\left[1 - \left(2\frac{(y - y_0)}{h_0}\right)^2\right]^{1/4} \tag{8.22}$$

$$c_2(0, y) = c_0(0, y_0)b_0^2(0, y_0)\left[1 - \left(2\frac{(y - y_0)}{h_0}\right)^2\right]^{3/4} \tag{8.23}$$

K_z is assumed to follow the $^{4/3}$ law,

$$k_z = Ab^{4/3} \tag{8.24}$$

where A lies in the range $(4 \times 10^{-4}$ to $3 \times 10^{-2})$ m$^{2/3}$ sec^{-1}. b is defined by $b^2 = c_2/c_0$ and $L = 4b$, where L is the width of the jet at x, y.

Assuming Gaussian profiles are maintained

$$c_0(0, y_0) = \int_{-\infty}^{\infty} c(0, y_0, z)\, dz$$

$$= \int_{-\infty}^{\infty} c(0, y_0, 0)\exp\left(\frac{-z^2}{b_0^2}\right) dz$$

i.e.

$$c_0(0, y_0) = c(0, y_0, 0)b_0(0, y_0)\sqrt{\pi} \tag{8.25}$$

$$c_2(0, y_0) = c(0, y_0, 0)b_0^3(0, y)\sqrt{\pi} \tag{8.26}$$

and

$$c(x, y, 0) = c(0, y, 0)\frac{c_0(x, y)}{c_0(0, y_0)}\frac{b_0(0, y_0)}{b(x, y)} \tag{8.27}$$

The equations (8.20) and (8.21) and hence equation (8.27) are solved using a finite difference scheme on two rectangular grids in the x, y domain with Δx (5 to 10 m) (Fig. 8.3).

c and U_a are defined at the grid points on the i rows and K_y on the j rows.

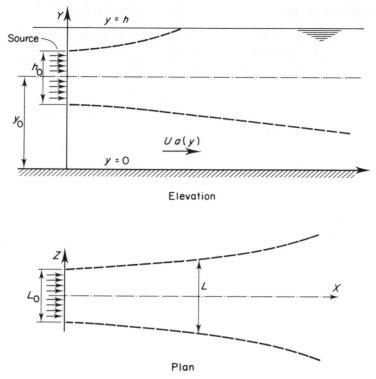

Fig. 8.3 Turbulent diffusion zone

Equation (8.20) is therefore represented as

$$U_i\left(\frac{c_i(x + \Delta x) - c_i(x)}{\Delta x}\right) = K_j\left(\frac{c_i + 1 - 2c_i + c_{i-1}}{\Delta y^2}\right)$$

$$+ \left(\frac{K_{j+1} - K_{j-1}}{2\Delta y}\right)\left(\frac{c_{i+1} - c_{i-1}}{2\Delta y}\right) - K_d c_i \qquad j = i, \ i = 2\ldots N \qquad (8.28)$$

where

$$c_i = c_0(x, \ y_i), \ u_i = U_a(x, \ y_i),$$
$$K_j = K_y(y_j)$$

The boundary conditions are

$$c_1 = c_2; \qquad c_N = c_{N+1}$$

Equation (8.21) is similarly represented.

An explicit numerical scheme is used to determine c_0 from the discretized form of equation (8.20), whence c_2 is similarly obtained from equation (8.21). Computations then proceed stepwise in intervals Δx.

8.4 BIBLIOGRAPHY

Abraham, G. and Brolsma, A. A. (1965). Diffusers for disposal of sewage in shallow tidal water, *Delft Hydraulics*. Lab. Pub. No. 37.

Fan, L.-N. (1967). Turbulent buoyant jets into stratified or flowing ambient fluids, (W. M. Keck), *Lab. of Hyd. and Water Resources*. Report No. KH-R-15, California Institute of Technology.

McBride, G. B. (1973). Numerical solutions of the equations governing submarine discharge of liquid waste, *Int. Conf. on Numerical Methods in Fluid Dynamics*. University of Southampton.

Rawn, A. M. and Palmer, H. K. (1930). Predetermining the extent of a sewage field in sea water, *Trans. ASCE* **94**, 1036–60.

APPENDIX

```
 10 REM BUOYANT JET MODEL
 15 G2 = 19.62
 20 DIMRO(3Ø), ZK(4,4)
 30 REM READ JETVEL, INCLINATION, DENSITY, DEPTH INTL
 40 READUI.T1.RI.DH
 50 PRINT"READING DS,DEPTH,OUTFALL DIA. ALPHA, LAMDA"
 60 READ SD, YØ,RD,AL,LA
 62 PRINT"READING AMBIENT DENSITY"
 63 K = YØ/DH + 1
 64 FORI = 1TOK
 65 READRO(I)
 66 NEXT
 67 DØ = RO(1) – RI
 70 REM FLOW ESTABLISHMENT
 71 FORI = 1TOK
 72 PRINTRO(I):NEXT
 80 T1 = T1*π/18Ø
 90 X = 6.2*RD*COS(T1)
100 Y = 6.2*RD*SIN(T1)
110 BI = RD/1.414
120 BO = BI:UO = UI
130 UB = UO*BO*BO
140 F = G2*LA*LA
150 G = F/2
160 D = (1 + LA↑2)/LA↑2 :PRINT"D = ";D
170 DO = 2/D
180 DØ = DØ*D/2
190 U = UI :B = BI :TH = T1 :DR = DØ
195 PRINTU,B,TH,DR
200 DR = DØ
```

```
210 KK = 1
220 PRINT"    VEL    WIDTH    THETA    DRO    X    Y    DILN"
225 DA = Ø
230 J = 1
240 C = 2
250 ZK(1,J) = SD*(F*DR*SIN(TH)/(U*RO(1))−2*U*AL/B)
260 ZK(2,J) = SD*(2*AL − B*G*SIN(TH)*DR)/(U↑2*RO(1))
270 ZK(3,J) = SD*(F*COS(TH)*DR)/(U↑2*RO(1))
280 ZK(4,J) = SD*(D*SIN(TH)*DA − 2*AL*DR/B)
290 IFJ = 4THEN460
300 U = UI + ZK(1,J)/C
310 B = BI + ZK(2,J)/C
320 TH = T1 + ZK(3,J)/C
330 DR = DØ + ZK(4,J)/C
340 DX = SD*COS(TH)
350 DY = SD*SIN(TH)
360 IFJ = 1THEN370
362 IFJ = 2THEN400
365 IFJ = 3THEN430
367 IFJ = 4THEN430
370 J = 2
380 C = 2
390 GOTO250
400 J = 3
410 C = 1:GOTO250
430 J = 4
440 C = 1:GOTO250
460 U = UI + (ZK(1,1) + 2*ZK(1,2) + 2*ZK(1,3) + ZK(1,4))/6
480 B = BI + (ZK(2,1) + 2*ZK(2,2) + 2*ZK(2,3) + ZK(2,4))/6
490 TH = T1 + (ZK(3,1) + 2*ZK(3,2) + 2*ZK(3,3) + ZK(3,4))/6
500 DR = DØ + (ZK(4,1) + 2*ZK(4,2) + 2*ZK(4,3) + ZK(4,4))/6
510 X = X + DX
520 Y = Y + DY
540 W = 2.828*B
550 DL = DO*U*B↑2/UB
555 PRINTKK
570 IFKK = 5ØTHEN590
588 GOTO610
590 PRINTU;W;TH;DR;X;Y;DL
595 KK = Ø
610 UI = U
620 BI = B
630 T1 = TH
```

```
 640  DO = DR
 650  JJ = Y/DH + 1
 660  DR = (RO(JJ + 1) − RO(JJ))/DY
 670  IFY > = YØTHEN710
 680  IFTH < ØTHEN710
 690  KK = KK + 1
 700  GOTO230
 710  END
2000  DATA2.0,0.0,1000.0,1
2010  DATA0.001,20,0.15,0.082,1.16
2020  DATA1025,1025,1025,1025,1025
2030  DATA1025,1025,1025,1025,1025
2040  DATA1025,1025,1025,1025,1025
2050  DATA1025,1025,1025,1025,1025
2060  DATA1025
READY
```

Chapter 9

Groundwater Quality Modelling

R.E. FEATHERSTONE

9.1 INTRODUCTION

The natural and, generally, high quality of ground water may be seriously affected by many types and sources of contamination associated with human activities and land use. Ground water pollution can arise from a variety of activites, notably

(a) the disposal of liquid effluent and sludge spreading directly on the out-crop of aquifers;
(b) disposal on soil from which it can percolate downwards to the aquifer;
(c) from leachate by percolating surface water from solid waste tips, by seepage from sewage treatment ponds;
(d) by sub-surface disposal of radio-active liquid wastes and
(e) due to saltwater intrusion in coastal aquifers.

Since groundwater flow systems are hydraulically connected with surface water systems, groundwater pollution can, in turn, lead to surface water pollution, for example, nitrate pollution due to the application of artificial fertilizers and salinity increases in alluvial stream-aquifer systems related to irrigation practices.

In recent years, increasing emphasis has been placed on groundwater quality and the development of methods to assist with the planning and design of projects to minimize groundwater contamination, and to simulate the transport of pollutants through aquifers to enable concentrations at discharge points such as abstraction wells, streams or springs to be estimated. The methods used incorporate both the theoretical description of the physical system together with field and laboratory studies to evaluate physical parameters related to the water and pollutant movements.

9.2 GOVERNING EQUATIONS OF GROUND WATER HYDRAULICS

Since convective transport and hydrodynamic dispersion depend on the seepage velocity of groundwater flow, the groundwater flow equation must be solved simultaneously with the mass transport equation describing the distribution of concentration of the pollutant. The groundwater flow equation is derived from the continuity equation in which the specific discharge is expressed by the Darcy linear seepage law

$$V_s = - K_s \, \partial h / \partial s \qquad (9.1)$$

where K_s is the coefficient of permeability for flow in the s direction and $\partial h / \partial s$ is the hydraulic gradient (slope of water table or piezometric surface). The following equation is obtained for two-dimensional flows:

$$\frac{\partial}{\partial x}\left(T_x \frac{\partial h}{\partial x}\right) + \frac{\partial}{\partial y}\left(T_y \frac{\partial h}{\partial y}\right) + q = S \frac{\partial h}{\partial t} \qquad (9.2)$$

where T_x and T_y are the transmissivities in the directions of the coordinate axes ($= Kb$), dimensions L^2/T;

  b is the depth of saturated flow, L;
  h is the hydraulic head in the aquifer, L;
  q is the vertical recharge, or volume flux, per unit area, L/T and
  S is the storage coefficient, L^0.

Equation (9.2) has been solved numerically by several methods including explicit and implicit finite difference methods and finite element methods. In each method the area is sub-divided by a grid into a number of smaller areas. Rectangular or square grids are commonly used in the finite difference methods.

Computations proceed in discrete time steps, values of $h_{x,y}$ being evaluated at the mesh points. The boundary conditions may be of two main types, the Dirichlet or fixed head boundary condition, and the Neumann boundary condition,

$$\partial h / \partial n = f(x, y),$$

where n is normal to the boundary.

The condition $\partial h / \partial n = 0$ describes an impermeable boundary. In using the two-dimensional equation (9.2) it is assumed that the velocity components V_x and V_y are uniform over the saturated depth, and that the flows are within the range of validity of the Darcy equation. In the case of a confined aquifer, the saturated thickness remains constant at each x,y as the head changes with time but in the water-table aquifer (unconfined aquifer) it varies with time. In the latter case equation (9.2) is expressed as

$$\frac{\partial}{\partial x}\left(K_x h \frac{\partial h}{\partial x}\right) + \frac{\partial}{\partial y}\left(K_y h \frac{\partial h}{\partial y}\right) + q = S \frac{\partial h}{\partial t}$$

which is non-linear. However, since this is computationally more demanding than the linear form (equation (9.2)), the latter is generally used for all aquifer types. A simple explicit finite difference form of equation (9.2) on a rectangular grid of size $\Delta x \times \Delta y$, and identifying the mesh points by x, y (i, j)

$$\frac{h_{ij}^{t + \Delta t} - h_{ij}^t}{\Delta t} = \frac{Tx}{S} \left[\frac{h_{i,j-1} - 2h_{ij} + h_{ij+1}}{\Delta x^2} \right]^t + \frac{Ty}{S} \left[\frac{h_{i-1,j} - 2h_{ij} + h_{i+1,j}}{\Delta y^2} \right]^t$$

Δt must be small to comply with the stability criterion,

$$\frac{\Delta t}{\Delta x^2} \frac{T}{S} < 0.25$$

A more commonly used finite difference method is the alternating direction implicit method in which the heads at each time step are evaluated simultaneously along each row. This is inherently stable and larger time steps than in the explicit method can be adopted. However, Rushton found that if sudden changes of head occur at a well very small time steps should be used.

The steady-state or equilibrium solution is simply obtained by writing $\partial h / \partial t = 0$ in equation (9.2).

9.3 ANALYTICAL SOLUTIONS

In the case of an isotropic, homogeneous aquifer, analytical solutions are possible for steady-state conditions. For example, the flow pattern in a uniform aquifer of infinite extent under a uniform hydraulic gradient in the region influenced by an abstraction well which is treated as a mathematical 'sink', can be described by the stream function

$$\psi_{r\theta} = Ur \sin \theta - q\theta / 2\pi$$

where U is the uniform specific discharge and q the well discharge per unit depth. The potential function, orthogonal to the stream function, is

$$\phi = Ur \cos \theta - \frac{q}{2\pi} \ell n \, r / r_w$$

where r_w is the radius of the well.
V_r is defined by $\partial \phi / \partial r$ and since $V_r = - K_r (\partial h / \partial r)$,

$$\frac{\partial \phi}{\partial r} = K_r \frac{\partial h}{\partial r}$$

whence

$$h_{x,y} - h_0 = x \left(\frac{dh}{dx} \right)_u + \frac{Q}{2\pi k b} \ell n \left(\frac{\sqrt{x^2 + y^2}}{r_w} \right)$$

in which h_0 = head of water in the abstraction well. (dh/dx_u) is the hydraulic gradient due to the uniform flow alone and Q $(= qb)$ the well discharge.

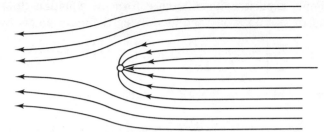

Fig. 9.1 Flow field near abstraction well in uniform ground water flow

Streamlines are orthogonal to the iso-head lines resulting in a typical flow pattern as shown in Fig. 9.1. Seepage, or pore velocity components, are given by

$$u_x = \frac{-K_x}{n} \frac{\partial h}{\partial x} \; ; \quad u_y = \frac{-K_y}{n} \frac{\partial h}{\partial y}$$

where n is the effective aquifer porosity (dimensionless).

9.4 MASS TRANSPORT

The two-dimensional convective-dispersion equation describing the depth-averaged concentration over the saturated thickness h may be written as

$$\frac{\partial C}{\partial t} + \frac{\partial}{\partial x_i} (Cu_{x_i}) - \frac{\partial}{\partial x_i} \left(D_{ij} \frac{\partial C}{\partial x_j} \right) - \epsilon C_0 = 0, \quad i,j = 1,2$$

where

x_i = horizontal coordinates $(i = i, 2)$, L
u_i = components of seepage velocity, LT^{-1};
ϵ = rate of inflow per unit area per unit depth T^{-1}
C = concentration of pollutant or tracer, ML^{-3}
C_0 = concentration of the inflow, ML^{-3}

D_{ij} = dispersion tensor $\begin{bmatrix} D_{11} & D_{12} \\ D_{21} & D_{22} \end{bmatrix}$, $L^2 T^{-1}$

The dispersion coefficients may be related to the seepage velocity of groundwater flow and the nature of the aquifer using a relationship due to Scheidegger,

$$D_{xx} = D_L \frac{(u_x)^2}{U^2} + D_T \frac{(u_y)^2}{U^2}$$

$$D_{yy} = D_T \frac{(u_x)^2}{U^2} + D_L \frac{(u_y)^2}{U^2}$$

$$D_{xy} = D_{yx} = (D_L - D_T) \frac{u_x u_y}{U^2}$$

The coefficients D_L and D_T are the longitudinal and lateral dispersion coefficients, respectively, defined by

$$D_L = \alpha_L U$$
$$D_T = \alpha_T U$$

where U is the magnitude of the resultant seepage velocity, α_L is the longitudinal dispersivity in the direction of flow and α_T is the transverse dispersivity normal to the direction of flow of an isotropic porous medium. Laboratory studies using homogeneous porous media show that D_L is normally greater than D_T by a factor of 5 to 20 and dispersivities lie in the range $10^{-2} - 1$ cm. Field studies yield values of dispersivity in the range $10 - 10^2$ m probably due to the non-homogeneous and anisotropic nature of aquifer structure.

A longitudinal dispersivity for sandstone of 0.6 m has been estimated by Oakes and Edworthy. Cole has reported a dispersivity of 12 m for chalk.

If the concentration of a pollutant in the leachate from the landfill is C_1 and the flow of groundwater is um/day the concentration of the pollutant below the landfill is

$$C_0 = \frac{\epsilon C_1}{\epsilon + ub/l}$$

where b is the saturated thickness of the aquifer and ϵ the mean percolation rate from the landfill area.

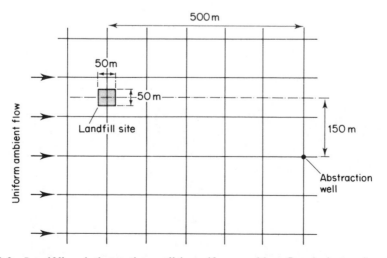

Fig. 9.2 Landfill and abstraction well in uniform ambient flow in isotropic aquifer

9.5 BIBLIOGRAPHY

Bear, J. (1972). *Dynamics of Fluids in Porous Media*. Elsevier, New York.

Cole, J.A. (1972). Some Interpretations of Dispersion Measurements in Aquifers In: *Groundwater Pollution in Europe*. Water Information Centre, New York.

Oakes, D.B. and Edworthy, K.J. (1976). Field measurements of dispersion coefficients in the UK, Conference on Groundwater Quality, Measurement, Prediction and Protection, Reading.

Ogata, A. (1970). Theory of dispersion in a granular medium, *U.S. Geol. Surv.* Prof. Pap. 411 – I, 1 – 34.

Remson, I., Hornberger, G.M. and Molz, F.J. (1974). *Numerical Methods in Subsurface Hydrology*. Wiley – Interscience, New York.

Rushton, K.R. and Tomlinson (1971). Digital computer solutions of groundwater flow, *Journal of Hydrology* **12**, 339 – 362.

Chapter 10

Modelling of Sedimentation

A. JAMES

10.1 INTRODUCTION

The physical process of removing suspended matter from wastewater by sedimentation has formed an essential part of treatment for many years. However, in spite of its long history, the prediction of the performance of sedimentation tanks has proved difficult especially when treating heterogeneous suspensions.

Two factors are mainly responsible for the variation in the way in which particles settle out of suspension

(a) the concentration of the suspension;
(b) the flocculating properties of the particles.

Figure 10.1 shows the broad classification of settlement into four main categories and the interaction of flocculation and concentration on the settling performance.

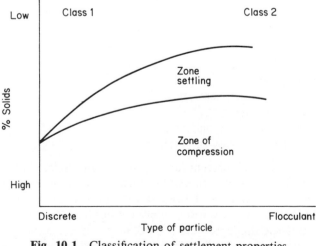

Fig. 10.1 Classification of settlement properties

169

Class 1 particles have little tendency to flocculate in dilute suspension whereas Class 2 particles tend to flocculate even in dilute suspension.

At low to intermediate concentrations of flocculant suspension the interparticle forces are sufficient to keep the particles in a fixed position relative to each other. As a result, the mass of particles subsides as whole in a regime described as zone settling. In this zone the rate of settling is controlled by the rate at which water passes upward through the mass and this is inversely related to the concentration. The particles are thus supported by the water between the particles and the upper particles are not supported by the lower ones.

In the compression zone, the concentration of particles is high enough for particles to come into contact with each other and each layer of particles provides mechanical support for the layers above it. The movement of particles is no longer governed only by hydraulic frictional forces and the links between particles. Forces transmitted by particle contacts via the compressible properties of the sludge and the interstitial pressure caused by the squeezing out of water from the compressing particles govern the settling behaviour in the compression zone.

The distinction between zone settling and compression is important in wastewater treatment because it is the rate of the former process (called clarification) which determines the design of primary sedimentation tanks and it is the rate of the latter process (called thickening) which determines the design of secondary sedimentation tanks.

10.2 MODELLING OF ZONE SETTLING

The settling of discrete non-flocculant particles may be described by Stokes Law

$$V_s = \frac{g\ d^{2(}S_s - 1)}{18v} \tag{10.1}$$

where

V_s = settling velocity
g = gravitational acceleration
d = particle diameter
S_s = specific gravity of particles
v = kinematic viscosity of water

but this only applies to quiescent laminar flow conditions. Under these conditions, the settling particles accelerate to a terminal settling velocity, V_s at which the effective mass is equal to the resistance of the water to sedimentation.

The determination of the settling velocity is complicated by the lack of symmetry in the particle shape. Since this is not uniform, it means that a coefficient of drag may be needed to reflect the change in drag resulting from variation in shape and skin friction. However, the effect of shape on settling velocity is unlikely to be significant at the velocities normally encountered in wastewater treatment. Further complications occur because flocculant particles do not have fixed shapes and tend to coalesce forming larger particles so that the settling velocity is no longer constant with time. Initially, the tendency to flocculate increases the rate of settling but at higher concentrations hindered settlement may occur. Agglomeration depends on the length of the settling path and the nature of the particle. This makes the general mathematical formulation of settling velocity a complex problem.

In these circumstances two modelling approaches have been adopted—deterministic and stochastic.

The deterministic approach is based on the concept of a plugflow reactor and non-flocculating discrete particles. Using these assumptions if a particle with settling velocity V_s enters a tank at the surface and is just settled at the outlet end then

$$V_s = \frac{Q}{A} \tag{10.2}$$

where

Q = flow
A = surface area of tank

In the case of a suspension with settling velocity u (where $u < V_s$) the proportion of solids that will settle out is given by u/V_s and

$$C_e = C_0 \left(1 - \frac{u}{V_s}\right) \tag{10.3}$$

where

C_e = concentration of solids in effluent
C_0 = concentration of solids in influent

If this equation is applied to a series of n basins all of equal volume then

$$C_e = C_0 \left(1 - n\frac{u}{V_s}\right)^n \tag{10.4}$$

When the basins are completely mixed reactors then

$$C_e = \frac{C_0}{[1 + n(u/V_s)]^n} \tag{10.5}$$

In the limit as $n \Rightarrow \infty$ both equations (10.4) and (10.5) tend to

$$C_e = C_0\, e^{-u/V_s} \tag{10.6}$$

The above expressions relate the removal of solids solely to the surface area but it has become apparent that the density currents and circulation in the tank require to be taken into account if realistic results are to be obtained from the model. This gives rather more complex expressions such as that given in equation (10.7)

$$C_e = \frac{C_0(1-r)}{1 - \left[\dfrac{B[1-(\phi^n/P)]}{P-\phi}\right] - n\left(\dfrac{\phi}{P}\right)^n} \tag{10.7}$$

where

$$B = \left(\frac{u}{V_s}\right)\bigg/\left(\frac{u}{V_s} + \frac{1}{1-r}\right)$$

$$P = 1 + \frac{u}{V_s}\left(\frac{1-r}{r_n}\right)$$

$$\phi = \left\{\frac{[1/(1-r)]}{[n(u/V_s) + 1/(1-r)]}\right\}$$

$$r = \frac{Q1}{Q + Q_1}$$

$$Q = \text{flow}$$

$$Q1 = \text{reverse current flow}$$

as $n \Rightarrow \infty$ then the concentration of solids in the effluent is given by equation (10.8)

$$C_e = \frac{C_0(1-r)\exp\left[-(u/V_s)(1-r)\right]}{1 - r\exp\left\{-(u/V_s)[(1-r^2/r]\right\}} \tag{10.8}$$

10.3 MODELLING OF SECONDARY SEDIMENTATION TANKS

The performance of secondary sedimentation tanks is governed by the overflow rate (clarification) and the transport of solids to the bottom of the tank (thickening). Either process may control the overall efficiency but in many secondary settling tanks thickening is clearly the critical process.

The modelling of thickening has been based on a mass flux approach. This is made up of two components:

(a) Gravity settlement.
(b) Advective movement due to the sludge removal.

The gravity settlement causes a solids flux which may be expressed as the product of the settling velocity and the solids concentration.

$$G_s = V_s X \tag{10.9}$$

G_s = solids flux due to gravity settlement

X = solids concentration

The advective transport may be expressed in terms of the draw-off rate Q_r, the cross-sectional area of the sedimentation tank A and the solids concentration, X, as shown in equation (10.10).

$$G_a = U_a X \qquad (10.10)$$

where

$$G_a = \text{advective flux}$$

and

$$U_a = \text{underflow velocity} = Q_r/A$$

The total solids flux, G_t is therefore the sum of these two components.

$$G_t = G_a + G_s$$
$$= X(U_a + V_s) \qquad (10.11)$$

This is represented graphically in Fig. 10.2. It shows that there is a minimum value of the flux G_L at a solids concentration X_L, which limits the rate at which solids reach the bottom of the sedimentation tank. To ensure that all the solids reach the base of the settling tank, the applied flux must not exceed this limiting flux, i.e.

$$G_L = X_0(Q_i + Q_r)/A \qquad (10.12)$$

$$G_L = X_0(U_0 + U_a) \qquad (10.13)$$

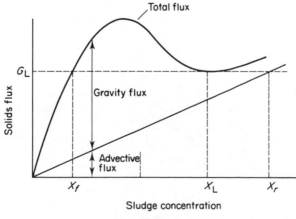

Fig. 10.2

where
  X_0 = mixed liquor suspended solids
  Q_i = influent flow
  G_L = limiting flux
  $U_0 = Q_i/A$

A mass balance on the solids in sedimentation tank as given by equation (10.14)

$$X_0(Q_i + Q_r) = X_r Q_r \qquad (10.14)$$

assuming no loss of solids in the effluent.

This may be modelled, as shown in Fig. 10.3 by selecting appropriate values for X_0 and Q_i having established by batch settling tests the relationship between V_s and X. This relationship is usually expressed by:

$$V_s = V_0\, e^{-nx} \qquad (10.15)$$

where V_0 and n are constants describing the settling characteristics of the sludge.

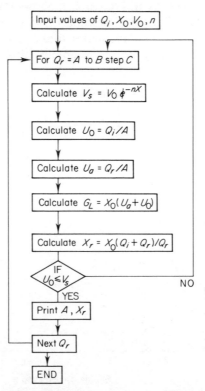

Fig. 10.3 Flow diagram for design model of thickening

The listing of the program for this simple model is given in Appendix 1 of this chapter.

An alternative approach is to consider the secondary sedimentation tank as part of the activated sludge process, where the thickened solids are recycled through the aeration tank and ultimately back to the settling tank. The recycle ratio may be expressed as a fraction r of the waste flow Q.

Assuming that the relationship between the settling velocity and the suspended solids may be represented as

$$V_s = aX_0^n \tag{10.16}$$

where a and n are constants, then substituting for V_s in the flux equation gives

$$G_t = aX_0^{(1-n)} + \frac{rQX_0}{A} \tag{10.17}$$

Differentiating with respect to X_0 gives

$$\frac{\partial G_t}{\partial X_0} = a(1-n)X_0^{1-n} + \frac{rQ}{A} \tag{10.18}$$

From Fig. 10.2 it can be seen that X_L occurs at $\partial G_t/\partial X_0 = 0$ which gives

$$X_L = \left[\frac{a(n-1)}{r} \cdot \frac{A}{Q}\right]^{1/n} \tag{10.19}$$

For G_L to be a minimum then

$$\frac{\partial^2 G_L}{\partial X_L^2} = \frac{an(n-1)}{X_0(1+n)} > 0 \quad \text{for positive values of } V_s, a \text{ and } n > 1 \tag{10.20}$$

Substituting for X_L into the flux equation gives the value of the limiting flux G_L in terms of Q, A, r, a and n.

$$G_L = \left[a(n-1)\right]^{1/n}\left[\frac{n}{n-1}\right]\left[r\frac{Q}{A}\right]^{(n-1)/n} \tag{10.21}$$

The recycle concentration X_r from a secondary settling tank of cross-sectional area A loaded at a rate G_L, can be calculated from a mass balance to be as shown in equation (10.22)

$$X_r = \frac{G_L A}{rQ} \tag{10.22}$$

As with clarification theory, the thickening model has practical limitations, namely:

(a) Establishing the relationship between the settling velocity and the concentration of solids.
(b) The determination of constants a and n in equation (10.16).

(c) The assumption that the velocity distribution is uniform over the cross-section of the tank.

(d) With highly compressible solids the settling velocity becomes a function of sludge depth.

10.4 COMPUTER-AIDED DESIGN OF SEDIMENTATION TANKS

Since the conventional design of primary sedimentation tanks is on such a simple basis, it can serve as an example of computer-aided design. Obviously, a similar approach could be used to the design of aeration tanks or other wastewater treatment units.

The example of CAD is given in the form of a program written in BASIC in Appendix 2 of this chapter. All the stages are identified in the program by comments (written as REM $-----$) but the following notes may be helpful in understanding the detailed working.

1. Process Design Stage—Lines 5–350
   The design is based upon the flow Q, the settling velocity of the particles $V1$ and a coefficient expressing the hydraulic efficiency of the tank, T. Having calculated the area required A (Line 30), the volume, V, is calculated on the basis of two hours retention (Line 40) and this is used to calculate the Depth, D (Line 60). For rectangular tanks the calculation of width, B is based upon the relationship

$$B = 0.4 \times \sqrt{\text{area}}$$

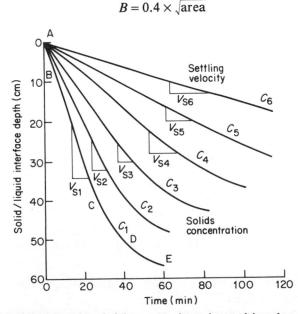

Fig. 10.4 Solid–liquid interface height versus time observed in a batch settling test

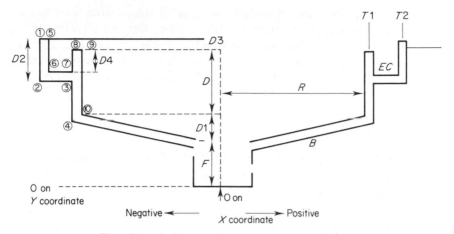

Fig. 10.5 Sedimentation tank graphical design

with the constraints that the square of the length, L divided by $B*D$ should not be greater than 20. If this is exceeded then B is reduced until the constraint is satisfied. For circular tanks the only dimension required is the radius which is calculated from the Volume/Depth (Line 200). The calculation of sludge storage (Lines 230–350) is based upon the slope of the base slab (4% for rectangular tanks and 8% for circular tanks).

2. Structural Design (Lines 500–1000)

   The ideas in this section are concerned with the use of reinforced concrete for water retaining structures. The aim is to avoid cracking by using slabs and walls of thickness Length/8 with a minimum thickness of 0.25 m.

3. Calculation of Costs and Quantities (Lines 1000–5000)

   The costs of concrete and steel, formwork, etc. are based upon simple price per volume relationships. Likewise, excavation quantities are based on the tank being set flush with the ground surface on a level site.

Table 10.1 Calculation of coordinates for CAD of a primary sedimentation tank

Point number	X Coordinate	Y Coordinate
1	$-20.9(R + T1 + EC + T2)$	$8.5(F + D + D1 + D3)$
2	$-20.9(R + T1 + EC + T2)$	$7.5(F + D + D1 + D3 - D2)$
3	$-20.2(R + T1)$	$7.5(F + D + D1 + D3 - D2)$
4	$-20.2(R + T1)$	$3.0(F + D1 - B)$
5	$(R + T1 + EC)$	$(F + D + D1 + D3)$
6	$(R + T1 + EC)$	$(F + D + D1 - D4)$
7	$(R + T1)$	$(F + D + D1 - D4)$
8	$(R + T1)$	$(F + D + D1)$
9	(R)	$(F + D + D1)$
10	(R)	$(F + D1)$

    The program does not include the graphic routes since these are particular to the type of computer being used. But the essentials of the preparation of drawings are illustrated by Table 10.1 and Fig. 10.5. The table and figure show the basis for calculating the coordinates and the graphics package simply contains the sub-routines for MOVE and DRAW.

BIBLIOGRAPHY

As indicated in Section 10.1 the modelling of sedimentation has not reached an advanced level. However the following publications will be useful background.

McKeinath, T. M. and Wanielista, M. (1975). Mathematical modelling for water pollution control processes, Ann Arbor Science, Ann Arbor, Mich. (Especially the chapter by DICK.)
Kynch, G. J. (1952). A Theory of Sedimentation, *Transactions of the Faraday Society.* **48**, 166.

APPENDIX 1

Listing of a model for the design of a secondary sedimentation tank

```
 10 READ Q1, X0, V0, N
 20 INPUT B, C, D, U0
 30 FOR Q2 = B to C STEP D
 40 V1 = V0 * (EXP (− N * X0))
    IF V1 < U0 GO TO 10
 50 A = Q1/U0
 60 U1 = Q2/A
 70 G = X0 * (U0 + U1)
 80 X1 = (X0 * (Q1 + Q2))/Q2
 90 PRINT A, X2
100 NEXT Q2
110 END
```

Notes

Q1 is the flow rate
X0 is the concentration of mixed liquor suspended solids
V0 is the settling velocity of the solids
N   is a constant describing the settling properties of the solids
U0 is the overflow rate
Q2 is the return sludge flow rate

U1 is the underflow velocity
G   is the limiting flux

APPENDIX 2

CAD program for a primary sedimentation tank

```
5      REM CALC OF DIMENSIONS
10     INPUT Q, V1
20     T = .9
30     A = 0/(V1*T)
40     V2 = 2*Q
50     D = V2/A
60     PRINT "DEPTH = ", D
70     PRINT "AREA = ", A
80     PRINT "VOLUME = ", V2
90     REM CHOICE OF SHAPE
100    PRINT "IS TANK CIRCULAR"
110    INPUT S
120    IF S = 1 THEN 200
130    B = . 4*SQR(A)
140    L = A/B
150    O = (L↑2)/(B*D)
160    IF C > 20 THEN 190
170    B = B − .5
180    GOTO 130
190    PRINT "LENGTH = ", L, "WIDTH = ", B
195    GOTO 220
200    R = SQR(V2/(22*(D/7)))
210    PRINT "RADIUS = ",R
220    REM CALC OF SLUDGE STORAGE
230    IF S = 1 THEN 300
240    V3 = .5*(L*L/25)
250    V4 = V2 + V3
260    D1 = L/25
270    D2 = D + D1
280    PRINT "TOTAL VOLUME = ", V4
290    PRINT "MAXIMUM DEPTH = ", D2
295    GOTO 500
300    V3 = .5*(2*R*(8*R/100))
310    V4 = V3 + V2
320    D1 = 8*R/100
```

```
330   D2 = D1 + D
340   PRINT "TOTAL VOLUME = ", V4
350   PRINT "MAXIMUM DEPTH = ", D2
500   REM STRUCTURAL DESIGN
505   REM VALC OF BASE SLAB
510   IF S = 1 THEN 590
520   T = L/8
530   IF T > .2 THEN 550
540   T = .2
550   V5 = T*L*B
560   V6 = 3*V5/100
570   PRINT "VOLUME OF CONCRETE = ", V5
580   PRINT "VOLUME OF STEEL = ", V6
585   GOTO 670
590   T = R/8
600   IF T > .2 THEN 620
610   T = .2
620   V5 = T*(22/7)*R↑2
630   V6 = V5*3/100
640   PRINT "VOLUME OF CONCRETE = ", V5
650   PRINT "VOLUME OF STEEL = ", V6
660   REM CALC OF WALLS
670   T1 = L/5
700   IF S = 1 THEN 760
705   PRINT "T1 = ", T1, D,B
710   V7 = T1*D*B
720   V8 = T1*D2*B
730   V9 = 2*T1*((D + D2/2)*L)
740   Y1 = V7 + V8 + V9
750   Y2 = Y1/25
755   GO TO 780
780   PRINT "VOLUME OF CONCRETE = ", Y1
790   PRINT "VOLUME OF STEEL = ", Y2
800   REM CALC OF EFFLUENT CHANNEL
810   IF S = 1 THEN 850
820   Y3 = .02*B
830   Y4 = Y3/25
840   GOTO 870
850   Y3 = .02*(B + .025)
860   Y4 = Y3/25
870   PRINT "VOLUME OF CONCRETE = ", Y3
880   PRINT "VOLUME OF STEEL = ", Y4
1000  REM CALC OF COST
```

```
1005   REM COST CONCRETE AND STEEL
1010   Y5 = V5 + Y1 + Y3
1020   C1 = Y5*20
1030   PRINT "COST OF CONCRETE = ", C1
1040   Y6 = V6 + Y2 + Y4
1050   C2 = Y6*50
1060   PRINT "COST OF STEEL = ", C2
1070   REM COST OF EXCAVATION
1080   C3 = V4*1.1*3
1085   PRINT "COST OF EXCAVATION", C3
1090   REM COST OF BLINDING
1100   IF S = 1 THEN 1140
1110   A2 = L*B
1120   C4 = A2
1130   GOTO 1160
1140   A2 = 22/7*(R + .025)
1150   C4 = A2
1160   PRINT "COST OF BLINDING = ", C4
1170   REM COST OF FORMWORK
1180   IF S = 1 THEN 1220
1190   A3 = (D*B) + (D2*B) + ((.5*(D + D2))*L
1200   C5 = A3*3
1210   GOTO 1240
1220   A3 = D*(44/7)*R
1230   C5 = A3*5
1240   PRINT "COST OF FORMWORK = ", C5
1250   REM CLACL OF TOTAL COST
1260   C6 = C1 + C2 + C3 + C4 + C5
1270   C7 = C6*1.3
1280   PRINT "TOTAL COST = ", C7
5000   END
```

Chapter 11

Activated Sludge Models

A. JAMES AND D. J ELLIOTT

11.1 INTRODUCTION

The activated sludge process is one alternative frequently used for the removal of organic material in wastewater. In this process a large population of micro-organisms is maintained in suspension in a tank through which wastewater passes. Air or oxygen is supplied and purification takes place in a series of steps in which bacteria utilize the organic material to yield new cells and provide energy. Synthesized cells (sludge) are removed by sedimentation in a separate tank to produce a clarified effluent. A proportion of the active bacteria (activated sludge) is recycled to the inlet of the aeration unit to maintain the active micro-organism population. The organisms are maintained in the aeration tank in the form of flocs which are dispersed throughout the liquor to improve the chance of contact with the waste organic material.

Purification is achieved as a direct result of micro-organism metabolism which depends on the presence of sufficient oxygen and the high contact rate between the activated sludge and the waste. The end result is that a portion of the waste is converted to inorganic material (CO_2, H_2O, NH_3) by oxidation which produces energy for synthesis in which the other part of the waste is converted to new cells. In addition to the conversion of carbonaceaous organic matter, nitrifying bacteria will convert ammonia to nitrates. The rate depends on the retention characteristics of the particular process.

The term 'activated' refers to the ability of the sludge floc to quickly absorb colloidal and suspended material from solution. Initial removal of organics is due almost entirely to adsorbtion. Synthesis is proportional to biological oxidation and acts at a slower rate than the adsorptive process. Hence, organics initially adsorbed but not immediately synthesized are stored in the activated sludge floc. When the full capacity of the sludge has been utilized it becomes inactive in the adsorptive sense and activity can only be restored by a period of aeration during which stored material is used in oxidation and synthesis.

The micro-organisms are referred to as mixed liquor suspended solids (MLSS) which in practice represents all suspended matter in the system both

182

inert and biological. An alternative way of expressing the active portion of the biological sludge is to use the volatile mixed liquor suspended solids which is a measure of the organic fraction and has a value approximately 80% of the MLSS.

11.2 PROCESS DESIGN AND OPERATION

Several types of activated sludge processes are currently employed in which wastewater is mixed and aerated with activated sludge for periods of between one and thirty hours. A diagram of a conventional system is shown in Fig. 11.1

The process is operated as a continuous rather than a batch system. The traditional aeration tank design was a nominal plug-flow system. Typical tank dimensions are 4 m deep, 8 m wide and 30 – 100 m long. Dye tracer studies have shown that they do not approximate stirred tank or plug-flow systems particularly well. Considerable back mixing does occur however and considerable concentration gradients do exist. Mixed liquor is retained in the aeration basin for a period of 4 to 8 hours after which it flows to a settling tank or clarifier. Clarified liquid is discharged at the top and the settled solids are removed as sludge from the base. A portion of the sludge equal in volume to between 10 and 30% of the process stream flow is returned to the influent of the aeration tank.

Efficient operation of the clarifier is essential to the effectiveness of the process as a whole. Essentially, the treatment process is the conversion of solid and dissolved organic material to cellular material in suspension which is separated from the bulk liquid in the clarifier. The condition of the cellular material depends the settling characteristics of the sludge and hence the efficiency of the treatment process.

Two parameters of interest in both design and operation of activated sludge plants are Sludge Loading Rate and Sludge Age. The Sludge Loading Rate is a measure of the food to micro-organism ratio and is represented by kg BOD per kg MLSS per day. An alternative way of expressing this parameter is to use the ratio of BOD applied to the weight of volatile suspended solids fraction of the activated sludge suspension. The choice of SLR established Sludge Quality, BOD reduction efficiency, Synthesis Rate giving the quantity of excess

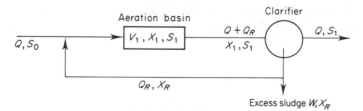

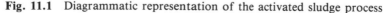

Fig. 11.1 Diagrammatic representation of the activated sludge process

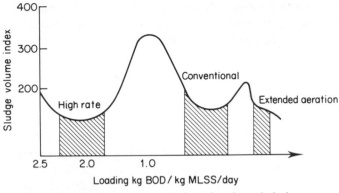

Fig. 11.2   Settling characteristics of activated sludge

sludge and the oxygen requirement. As mentioned earlier the most important consideration is the effect on sludge quality which may be measured by the Sludge Volume Index. The SVI is the volume occupied by gram of sludge in a 1-litre cylinder after 30 minutes quiescent settling. The higher the SVI the poorer the sludge quality in terms of its settling characteristics. A figure of SVI < 100 is considered a desirable value. Figure 11.2 relates sludge quality to sludge loading rate and defines operating ranges for the three important varieties of the activated sludge process.

Sludge age or biological solids retention time is a second parameter relating to the process variant. It is defined as the total active solids in the system divided by the total quantity of solids wasted per unit of time.

(a) *Extended aeration*

This variant of the activated sludge process has aeration tanks designed for long hydraulic retention of approximately 20 hours. All the sludge from the clarifier is returned to the aeration basin. As a result, the sludge concentration is high in the aeration basin and oxidation takes place to the extent that most cellular material is decomposed to inorganic forms. The only solids wasted are those which are carried over in the clarified effluent from the settling tank. The sludge age is therefore high and the sludge loading rate is low as shown in Fig. 11.2. BOD reduction achieved by extended aeration is up to 98% suitable in situations when little dilution water is available. The effluent is well nitrified, sludge disposal costs are low but large aeration volume and aeration power are needed.

(b) *High-rate process*

The high-rate process achieves 60 to 70% BOD reduction and produces more sludge than the conventional or extended aeration process. Between 5 and 10%

of the waste flow is returned as activated sludge and the aeration time is limited to 2 to 3 hours. The SLR is high giving a high growth rate but the sludge is difficult to settle producing a poor quality final effluent.

(c) *Choice of process variant*

The conventional system represents the best compromise between efficiency of treatment, capital and operating cost, operation and sludge disposal over the widest flow range. If 8 : 1 dilution is available the effluent will satisfy most consent standards. There is also a good economic balance between aeration facilities and sludge treatment and disposal. Power and nutrient costs are not excessive however the process requires considerable operating control and can be upset rather easily. It is also slow to return to normal efficiency after a perturbation in the system. For these reasons completely mixed processes were introduced during the 1950's which were found to have greater stability whilst retaining similar effluent quality characteristics to nominal plug-flow systems of the same volume. Considerable dilution is provided in the completely mixed system which helps prevent upset to the system from shock loadings of toxic materials or degradable organics. This system also provides a more uniform biological environment for the organisms which may be one reason for the improved system stability.

11.3 MODEL STRUCTURE

The initial ideas came from Monod's theory of the relationship between growth rate and substrate concentration which together with the concept of yield gave for a completely stirred reactor the coupled differential equations:

$$\frac{dX_2}{dt} = \hat{\mu}\left(\frac{S_2}{K_s + S_2}\right)X_2 - \frac{Q}{V}X_2 \qquad (11.1)$$

$$\frac{dS_2}{dt} = \frac{Q}{V}S_1 - \frac{Q}{V}S_2 - \frac{\hat{\mu}}{Y}\left(\frac{S_2}{K_s + S_2}\right)X_2 \qquad (11.2)$$

These equations need to be modified to take account of the recycle of bacteria which is the special feature of the activated sludge process. These become

$$\frac{dX_2}{dt} = \frac{Q}{V}X_1 - \left(\frac{(Q + Q_r)}{V}\right)X_2 + \frac{Q_r}{V}X_r + \hat{\mu}\left(\frac{S_2}{K_s + S_2}\right)X_2 - K_d X_2 \qquad (11.3)$$

$$\frac{dS_2}{dt} = \frac{Q}{V}S_1 - \left(\frac{(Q + Q_r)}{V}\right)S_2 + \frac{Q_r}{V}S_2 - \frac{\hat{\mu}}{Y}\left(\frac{S_2}{K_s + S_2}\right)X_2 \qquad (11.4)$$

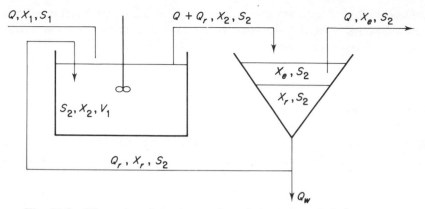

Fig. 11.3   Diagrammatic representation of the activated sludge process

where the symbols are as shown in Fig. 11.3. The use of the recycle overcomes the limitations imposed by the growth rate of the bacteria. i.e.

$$\frac{Q}{V} \quad \text{may be greater than } \mu$$

This is important since the K_s value for domestic waste is often reported as around 60 mg 1^{-1} of BOD and therefore with a need for an effluent BOD of 20 the value of μ becomes

$$\hat{\mu}\left(\frac{20}{20 + 60}\right) = 0.25\,\hat{\mu}$$

This together with the low ambient temperature and low viability gives μ values for activated sludge in the range of $0.6-2.0$ per day.

   Recycling also separates the retention time of the bacteria from that of the liquid. The latter is still given by V_T/Q where V_T is the total volume of the aeration and sedimentation tanks. The retention time of the bacteria is usually referred to sludge age. This is generally approximated as the quantity of bacteria stored in the aeration tank divided by the rate of loss of bacteria in the effluent and in the waste sludge.

$$\theta = \frac{V_A X_2}{(Q_w X_r + (Q - Q_w)X_2)} \tag{11.5}$$

where

$$\theta \;\; = \text{sludge age}$$
$$V_A = \text{volume of aeration tank}$$

Sludge age is important because the settling properties of the bacterial flocs change significantly with their retention time in the system as shown in Fig. 11.2.

The concept of sludge age can be used in calculating the steady-state solutions for $\bar{X}_2$ and $\bar{S}_2$. These are as follows:

$$\bar{S}_2 = \frac{K_s(1 + K_d\theta)}{\theta(Y - K_d) - 1} \tag{11.6}$$

$$\bar{X}_2 = \frac{Y(S_1 - S_2)}{1 + K_d\theta} \tag{11.7}$$

Since treatment units are intended to operate at quasi-steady state these equations can be useful to give a first approximation of the values of effluent BOD ($\bar{S}_2$) and MLSS ($\bar{X}_2$).

These equations are also useful for setting the initial conditions when using a dynamic model. In the absence of other data the steady state values for X_2 and S_2 can be used as the starting points for the simulation.

The above equations may be used in two ways:

(a) In the design mode.
(b) In the simulation mode.

For the first purpose the equations have been rewritten to give

$$U = \frac{Q}{V}\left(\frac{S_1 - S_2}{X_2}\right) \tag{11.8}$$

$$\frac{1}{\theta} = YU - Kd \tag{11.9}$$

where

U = Specific rate of substrate utilization

and can also be used in the design mode to determine the minimum retention below which bacteria are washed out of the system. At this point $S_2 = S_1$ and therefore

$$\frac{1}{\theta_c} = \frac{\hat{\mu}S_1}{S_1 + K_s} - K_d \tag{11.10}$$

where θ_c = critical retention time for solids.

Taking a mass balance on substrate across the reactor only gives

$$\frac{Q}{V}S_1 + \frac{Q_r}{V}S_2 - \frac{\mu}{Y}X_2 = (Q + Q_r)S_2 + \frac{dS_2}{dt} \tag{11.11}$$

This equation assumes that the substrate concentration in the recycled flow is the same as that in the reactor.

The steady-state form of this equation may be rearranged to give the total biomass of bacteria in the reactor as

$$VX_2 = Q(S_1 - S_2)\frac{Y}{\mu} \tag{11.12}$$

substituting for $\mu = 1/\theta + K_d$ this becomes

$$X_2V = \frac{YQ\theta(S_1 - S_2)}{1 + K_d\theta} \tag{11.13}$$

A bacterial balance across the aeration tank gives

$$Q_rX_r + (\mu - K_d)X_2V = (Q + Q_r)X_2 + V\frac{dX_2}{dt} \tag{11.14}$$

By substituting for $\mu = 1/\theta + K_d$ this becomes

$$Q_rX_r + \frac{1}{\theta}X_2V = (Q + Q_r)X_2 \tag{11.15}$$

Dividing through by X_2V and rearranging gives

$$\frac{1}{\theta} = -\frac{Q_rX_r}{X_2} + \frac{(Q + Q_r)}{V} = \frac{1}{V}\left(Q + Q_r - \frac{Q_rX_r}{X_2}\right) \tag{11.16}$$

or

$$\frac{1}{\theta} = \frac{Q}{V}\left(1 + \frac{Q_r}{Q} - \frac{Q_rX_r}{QX_2}\right) \tag{11.17}$$

In the design it is necessary to have prior knowledge of the coefficients K_s, K_d, μ and Y for the substrate and bacteria concerned. Given this information the following procedure may be used to develop an initial design.

1. Determine the required substrate concentration in the effluent, S_2 from the consent conditions.
2. Use the equation

$$\frac{1}{\theta} = \hat{\mu}\left(\frac{S_2}{K_s + S_2}\right) - K_d$$

   to calculate the sludge age.
3. Use θ together with the design flow Q to calculate X_2v from the following equation

$$X_2V = \frac{YQ\theta(S_1 - S_2)}{1 + K_d\theta}$$

4. Determine X_r the maximum solids concentration in the recycle by experiment
5. Using equation (11.16)

$$\frac{1}{\theta} = \frac{Q}{V}\ 1 + \frac{Q_r}{Q} - \frac{Q_rX_r}{QX_2}$$

   determine the volume of the aeration tank for different values of the recycle ratio Q_r/Q.

An alternative design procedure would be to choose a value for V and then to calculate X_2 and Q_r.

Having established a design from steady-state considerations the model may be used in the simulation mode for process control. This requires the simultaneous solution of the two mass balance equations

$$QS_1 + Q_rS_2 - \frac{\mu}{Y} X_2 V = (Q + Q_r)S_2 + V \frac{dS_2}{dt} \tag{11.18}$$

$$Q_rX_r + (\mu - K_d)X_2 V = (Q + Q_r)X_2 + V \frac{dX_2}{dt} \tag{11.19}$$

over time, starting with initial conditions obtained from the steady-state solutions and using arbitrary changes in the inputs so that the effect on the output can be compared with the required operating conditions.

The above equations are usually solved numerically using simple Euler or Runge – Katta routines. An example of a program incorporating both the design and simulation modes is given in the Appendix. This uses simple Euler and therefore requires a small time step to achieve numerical stability.

11.4 MODIFICATIONS TO THE MODEL

The model described in equation (11.2) is incomplete because it does not explicitly consider interactions between the biological reactor and the sedimentation process. The ratio of bacteria in the overflow X_e to the bacteria in the recycle X_r depends upon the sludge age but in many models it is incorporated as a fixed ratio. Where it is desired to make the model more realistic, the separation into X_e and X_r can be related to the sludge age using an arbitrary function describing Fig. 11.2.

Another aspect of the activated sludge process which is sometimes neglected in simulation models is the oxygen demand due to nitrification. Although this is a two-stage process

$$NH_3 \overset{K_1}{\rightarrow} NO_2 \overset{K_2}{\rightarrow} NO_3$$

since $K_2 > K_1$ the simulation need only represent the kinetics of the first stage.

Because the growth rate of the nitrifying bacteria is slower than that of carbonaceous bacteria, there may be a problem in activated sludge of maintaining sufficient nitrifying organisms. This may be achieved by manipulating the conditions in the aeration tank using the substrate concentrations to compensate for their different growth rates e.g.

for nitrosomonas $\mu = 0.25$ per day $K_s = 1$ mg 1^{-1} as $NH_3 - N$
for heterotrophs $\mu = 0.8$ per day $K_s = 60$ mg 1^{-1} as BOD

If substrate levels in the aeration tank are maintained at $NH_3N = 4$ mg 1^{-1} then

$$\text{nitrifying bacteria } \mu = 0.25\left(\frac{4}{1+4}\right) = 0.2$$

$$\text{heterotrophic bacteria } \mu = 0.8\left(\frac{20}{20+60}\right) = 0.2$$

Simulation models can be useful for examining the combinations of conditions necessary for obtaining nitrification. The model given in the Appendix includes the appropriate kinetic equations and coefficients but it should be noted that the effect of temperature on nitrifiers is more pronounced than on heterotrophs, so temperature correction may be necessary.

Another additional feature in some simulation models is the division of substrate into slowly degrading and rapidly degrading. In the most sophisticated models this division may be further extended to cover the following:

(a) Particulate or soluble,
(b) Biodegradable or non-degradable,
(c) Readily biodegradable or slowly biodegradable, which are all fractions of the influent COD. Such sophistication of the model can only be justified when data for evaluation the additional coefficients are available.

There are some fundamental objections to models based on the Monod kinetics as outlined above. The most serious objection is to the idea of using a fixed value for the Yield because the reaction between bacteria, organic matter and oxygen is progressive with time as shown in Fig. 11.4 so that the Yield clearly varies significantly.

When tested experimentally, it has been found that the Yield is only constant for a fixed sludge age and that the Yield decreased exponentially with in-

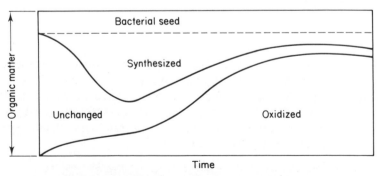

Fig. 11.4 Assimilation of substrate with time

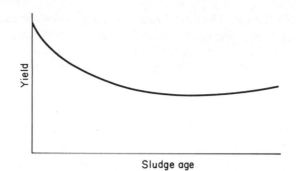

Fig. 11.5 Variation of yield with bacterial retention time

creasing sludge age. Various ideas have been suggested for overcoming this problem, without uncoupling the connection between substrate kinetics and bacterial kinetics. The simplest way is to allow the Yield to vary with sludge age and since the solids retention time is related to growth rate the relationship becomes

$$Y_0 = \frac{\mu Y}{\mu + mY} \tag{11.20}$$

where

Y_0 = observed yield at growth rate μ
m = maintenance energy

This has the advantage of requiring only minor changes in the conventional simulation models. Other suggestions for changes in the Monod kinetics have

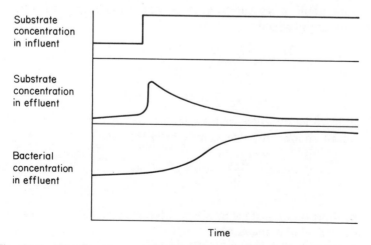

Fig. 11.6 Kinetic response of a laboratory-scale activated sludge system to a step increase in influent concentration

been rather more radical. They are based on the observation that there is a time lag in the system showing that growth is related to the concentration of stored substrate and not the concentration in the liquid. This is illustrated in Fig. 11.6. A more realistic simulation is therefore obtained by a two-step process

$$\text{Substrate} \rightarrow \text{Storage} \rightarrow \text{Metabolism}$$

The rate of substrate utilization is related to the difference between the concentration of storage products and the saturation storage capacity giving the equation

$$-\frac{dX_1}{dt} = a_1 K_1 \left(\frac{X_1}{X_s}\right)\left(S_T - S\right)$$

$$= a_1 K_1 \left(\frac{X_1}{X_s}\right)\left(\frac{S_T}{M} - \frac{S}{M}\right) M \qquad (11.21)$$

where

t = aeration time

X_1 = concentration of extra-cellular substrate

X_s = threshold concentration of X_1 below which X_1 is related to the ingestion rate

a_1 = conversion coefficient (substrate X_1/storage product S)

K_1 = ingestion rate coefficient

S = concentration of storage products

S_T = saturation storage capacity

M = concentration of cells

The rate of metabolism is assumed to be first order with respect to the concentration of storage products.

$$\frac{dS}{dt} = K_1 \left(\frac{X_1}{X_s}\right)\left(\frac{S_T}{M} - \frac{S}{M}\right) M - K_2 \left(\frac{S}{M}\right) M \qquad (11.22)$$

where

K_2 = rate coefficient for metabolism

The rate of degradation of the cells is represented by a first-order equation so that the overall change in bacteria may be represented by

$$\frac{dM}{dt} = \alpha a_2 K_s S - K_3 M \qquad (11.23)$$

where

α = proportion of storage products which is consumed for the synthesis of the cellular material

a_2 = conversion coefficient (M/S)

K_3 = death rate

The variation in the concentration of intermediate and end products is described by

$$\frac{dP}{dt} = a_4 K_3 M \tag{11.24}$$

where

P = concentration of end products
a_4 = conversion coefficient (P/M)

Variations in the activity of the bacteria may also be taken into account in the model by assuming that the rate of removal of substrate is influenced by the concentration of storage products. This may be defined by

$$\lambda = 1 - \frac{S}{S_T} \tag{11.25}$$

Substitution of the activity coefficient, λ into the above equation may be easily carried out, e.g.

$$\frac{d(1-\lambda)M}{dt} = K_1 \left[\frac{X_1}{X_s}\right] \lambda M - K_2 (1-\lambda)M \tag{11.26}$$

Although this much more detailed approach is a closer representation of the real situation, it does considerably increase the complexity of the modelling by introducing a whole new range of coefficients. The determination of the coefficients to be used in the Monod kinetic models has proved to be somewhat difficult and has led to wide variations in the published values. It therefore seems unlikely that models involving more coefficients will prove popular until more data become available.

11.5 BIBLIOGRAPHY

The modelling of activated sludge has attracted more attention than for all other forms of wastewater treatment. The following are some suggestions from a large literature.

Dold, P.L., Ekama, G.A. and Marais, G. v. R. (1980). A general model for the activated sludge process, *Progress in Water Technology.* **12**, 47–77.
Nyns, E.J. and Sheers, Y. (1977). Mathematical modelling and economic optimization of wastewater treatment plants, *Critical Reviews in Environmental Control.* **8**(1) pp. 1–89.
Pirt, S. J. (1965). The maintenance energy of bacteria in growing cultures, *Proceedings of the Royal Society.* Series B **163**, 224–231.
Tsuno, H., Goda, T. and Somiya, I. (1978). Kinetic model of activated sludge metabolism and its application to the response to qualitative shock loads, *Water Research.* **12**, 513–519.

Also appropriate Chapters in:

McKeinath, T.M. and Wanielista, M. (1975). *Mathematical modelling for water pollution control processes.* Ann Arbor Science. Ann Arbor, Mich.

James, A. (1978). *Mathematical Models in Water Pollution Control.* J. Wiley & Sons, Chichester.

APPENDIX

Activated sludge model

```
READY
10 REM ACTIVATED SLUDGE PROGRAMME
20 REM Q = WASTE FLOW (M3/HR)
30 REM Q1 = RETURN SLUDGE FLOW (M3/HR)
40 REM Q2 = WASTE SLUDGE FLOW (M3/HR)
50 REM X0 = CONC OF BACT IN  INF (MG/L)
60 REM X1 = CONC OF BACT IN MLSS (MG/L)
70 REM X2 = CONC OF BACT IN RETURN SLUDGE (MG/L)
80 REM X3 = CONC OF BACTERIA IN EFF (MG/L)
90 REM C0 = BOD OF INFLUENT (MG/L)
100 REM C1 = BOD OF MIXED LIQUOR (MG/L)
110 REM U = MAX GROWTH RATE(PER HOUR)
120 REM Y = YIELD
130 REM K1 = KS (MG/L)
140 REM K2 = KD (PER HOUR)
150 REM V = VOLUME OF AERATION TAN (M3)
160 REM T1 = LIQUID RETENTION TIME (HR)
170 REM T2 = SLUDGE AGE (HR)
500 READ Q,Q1,Q2,X1,X2,C0
510 READ C1,U,Y,K1,K2,V
520 READ B1,B2,K3,K4,U1,Y1,N1
550 DATA 40,20,2,1200,9000,250
560 DATA 20.,05.,5,60.,002,500
570 DATA 200,800,1,.0005,.02,.2,10
600 FOR H = 1 TO 10
1000 REM
1010 REM GROWTH AND UTILISATION MODEL
1020 REM G1 = RATE OF BACTERIAL CHANGE
1030 REM G2 = RATE OF SUBSTRATE CHANGE
1040 REM G3 = RATE OF NITRIFIER GROWTH
1050 REM G4 = RATE OF AMMONIA OXIDATION
1060 REM B1 = NITROSOMONAS IN MLSS
1070 REM B2 = NITROSOMONAS IN RETURN SLUDGE
1000 REM N0 = AMMONIA IN INFLUENT
```

```
1085 REM N1 = AMMONIA IN MIXED LIQUOR
1090 REM K3 = KS FOR NITROSOMONAS
1095 REM K4 = DEATH RATE FOR NITROSOMONAS
1100 G1 = Q1/V*X2 + X1*U*(C1/(K1 + C1)) – K2*X1 – (Q + Q1)/V*X1
1110 PRINT"G1 = ",G1
1200 G2 = Q/V*C0 + Q1/V*C1 – (Q + Q1)/V*C1 – X1*U/Y*(C1/(C1 + K1))
1210 PRINT"G2 = ",G2
1330 Z = (B1*U1)*(N1/(N1 + K3))
1331 Y = (B1*U1)*(N1/(N1 + K3))
1332 P = K4*B1333
1333 W = ((Q + Q1)*B1)/V
1334 G3 = Z + Y – P – W
1340 PRINT"G3 = ";G3
1360 G4 = Q/V*N0 + Q1/V*N1 – (Q + Q1)/V*N1 – B1*U1/Y1*(N1/(K3 + N1))
2000 REM SEDIMENTATION MODEL
2010 REM F = CONCENTRATION FACTOR = X2/X1
2100 T2 = X1*V/((Q – Q2)*X3 + Q2*X2)
2110 IF T2 < 75 THEN 2200
2120 IF T2 > = 75 < 120 THEN 2230
2130 IF T2 > = 120 THEN 2260
2200 F = 3.5
2210 GOTO 2300
2230 F = 2
2240 GOTO 2300
2268 F = 4 + .O1*T
2300 X2 = F*X1
2350 IF X2*(Q1 + Q2) > X1*(Q + Q1) THEN2450
2400 GOTO 2500
2450 X2 = ((Q + Q1)*X1)/(Q1 + Q2)
2460 F = X2/X1
2500 B2 = F*B1
2800 L = (Q + Q1)*X1 – (Q1 + Q2)*X2
2810 X3 = L/(Q – Q2)
2820 IF X3 < 0 THEN X3 = 0
3000 REM AERATION MODEL
3010 REM R1 = AERATION FOR BOD REMOVAL
3020 REM R2 = ENDOGENOUS RESPIRATION
3030 REM K5 = ENDOGENOUS RESP COEFF
3100 R1 = Q*(C0 – C1)
3200 R2 = V*X1*K5
3300 R3 = (R1 + R2)*3
3800 X1 = X1 + G1
3810 C1 = C1 + G2
```

```
3820  B1 = B1 + G3
3830  N1 = N1 + G4
3900  PRINT"MLSS = ",X1
3910  PRINT"EFF BOD = ",C1
3920  PRINT"SLUDGE AGE = ",T2
3930  PRINT"AERATION REQUIREMENT = ",R3
3940  PRINT"EFFLUENT SOLIDS CONC = "X3
4900  NEXT H
READY
```

Chapter 12

Modelling of Fixed Film Reactors

A. JAMES

12.1 INTRODUCTION

Fixed-film reactors have been used extensively for over 100 years in the treatment of wastewaters with the development of contact beds and percolating filters from land treatment. More recently the development of plastic packing materials has extended the scope of fixed-film reactors and has led to high-rate filters.

There is a marked contrast between the modelling of fixed-film reactors and the modelling of dispersed growth reactors. There is a large measure of agreement about the approach to modelling aeration tanks and many of the ideas have been incorporated into standard design procedures. Unfortunately, modelling of fixed-film reactors has not reached the same stage of development. For high-rate filters there is some measure of agreement about the approach to mathematical representaton and some limited success has been achieved in design but with percolating filters there is not even agreement about the concepts to be modelled.

Although the two types—percolating filters and high-rate filters are both fixed-film reactors, there are sufficient fundamental differences to make these distinct processes. The obvious differences are the much higher hydraulic and organic loading rates used in high-rate filters and the much greater use of recirculation. Also the use of plastic media with larger surface area and higher percentage voids. Less obvious but equally fundamental are the biological difference caused by these environmental changes. The mechanism for controlling the film growth in high-rate filters is hydraulic rather than biological: higher void ratios and thinner bacterial films make the flow regime independent of film thickness and higher organic transfer rates introduce the possibility of oxygen limitation.

The following section discusses the theory of organic and oxygen transfer into fixed films and the consequent film growth, followed by a section dealing with the modelling of high-rate filters and a final section dealing with percolating filters.

12.2 FIXED-FILM KINETICS

Various mechanisms have been postulated as controlling the rate of transfer of organic matter from a liquid layer passing over a static bacterial film which is attached to some inert support surface. The possibilities are represented diagrammatically in Fig. 12.1. The support surface is assumed to be inert and apart from its characteristics of specific surface and void ratio does not play any part in the process. Organic transfer in the two other layers can be considered separately.

The liquid layer has generally been regarded as flowing in a laminar manner over the bacterial film as shown in Fig. 12.2. The thickness of the liquid film d may be determined from the Nusselt equation

$$d = \left[\frac{3UQ}{sqW \sin \theta} \right]^{1/3} \tag{12.1}$$

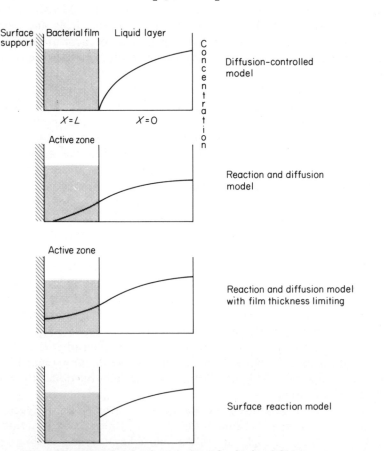

Fig. 12.1   Models of substrate transfer in fixed-film reactors

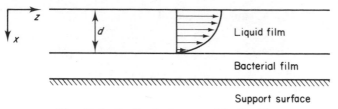

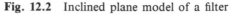

Fig. 12.2 Inclined plane model of a filter

where

U = average velocity
Q = flow
s = specific gravity of liquid
q = gravitational constant
W = width of the plane
θ = inclination of the plane to the horizontal

At any point within the liquid film at depth x the velocity U is given by

$$U = U_{max}\left[1 - \frac{x^2}{d}\right] \tag{12.2}$$

In such a laminar flow regime, transport at right angles to the streamlines can only occur due to molecular diffusion. The flux towards the bacterial film is described by Ficks' Law as shown in equation (12.3).

$$N_x = -D_L \frac{dc}{dx} \tag{12.3}$$

where

N_x = flux in x direction
D_L = diffusion coefficient in the liquid film

Mass transport along the streamlines N_z is given by the sum of the convective and diffusive transport components:

$$N_z = \mu c - D_L \frac{dc}{dz} \tag{12.4}$$

Assuming that there is no loss of organic matter within the liquid film, a mass balance on the unit element shown in Fig. 12.2 Δx, Δz is given by:

$$N_z\Big|_z x\Delta x - N_z\Big|_{z+\Delta z} \Delta x + N_x\Big|_x x\Delta z - N_x\Big|_{x+\Delta x} x \Delta z = 0$$

Dividing by Δx times Δz and taking the limits as Δx and $\Delta z \Rightarrow 0$ gives:

$$-\frac{dN_z}{dz} - \frac{dN_x}{dx} = 0$$

Substituting for N_x, N_z and U gives:

$$- U_{max} \left[1 - \left(\frac{x}{d} \right)^2 \right] \frac{dc}{dz} + D_L \frac{d^2 c}{dz^2} + D_L \frac{d^2 c}{dx^2} = 0 \qquad (12.5)$$

The longitudinal diffusive transport is relatively insignificant so this simplifies to equation (12.6).

$$U_{max} \left[1 - \left(\frac{x}{d} \right)^2 \right] \frac{dc}{dz} = D_L \frac{d^2 c}{dx^2} \qquad (12.6)$$

With the initial condition that no concentration gradient exists at the top of the plane, i.e.

$$c = c \quad \text{for all values of } x \text{ at } z = 0$$

and the boundary conditions that no transfer occurs at the air/liquid interface.

$$D_L \frac{dc}{dx} = 0 \text{ at } x = 0 \quad \text{for all values of } z$$

and that the flux at the liquid/bacteria interface is constant i.e.

$$- D_L \frac{dc}{dx} = N \text{ at } x = d \text{ for all values of } z$$

Other approaches have been suggested for this boundary condition such as:

$$N = \frac{K_1 L C^*}{1 + K_2 C^*} \qquad (12.7)$$

where

$$K_1 = a \frac{\alpha}{\beta}$$

$$K_2 = \frac{1}{\beta}$$

L = slime thickness
C^* = concentration of organic matter at liquid/bacteria interface

α and β are constants.

Or where the rate of diffusion in the liquid film is the overall rate controlling step then the concentration at the boundary may be assumed to be zero, i.e.

$$C = 0 \text{ at } x = d \quad \text{for all values of } z$$

In modelling the removal of organic matter within the bacterial film, it is necessary to consider the effect of film thickness. As indicated in Fig. 12.1 with thicker films there is an active zone and further increase in thickness does not alter the flux. In this case, the removal rate may be represented by

Michaelis–Menten kinetics as

$$N = K_A \frac{C^*}{K_s + C^*} \qquad (12.8)$$

where

$$K_A = \text{removal coefficient per unit area}$$

and

$$C^* = \text{concentration at interface}$$

This has also been described by a first-order reaction occuring at the surface, so that the flux becomes directly proportional to the remaining organic concentration

$$N = K_A[C^*] \qquad (12.9)$$

These two equations give similar fluxes for conditions where $K_s \simeq C$ as shown in Fig. 12.3 but departs at concentrations where $C \gg K_s$. Since the K_s values for filter slimes have been estimated in the range 120–180 mg l^{-1} both equations will give the same flux for normal operating conditions.

In filters where the slime layer is less than 50–70 μm thick, there is a finite concentration of organic matter at the bacteria/support surface interface. Under these conditions the metabolic flux of organic matter becomes proportional also to the bacterial numbers and is generally represented by multiplying the above equations by the film thickness, i.e.

$$N = K_A \left(\frac{C^*}{K_s + C^*} \right) x \qquad (12.10)$$

or

$$N = K_A[C^*] x$$

for $x =$ film thickness in the range 0–50 μm.

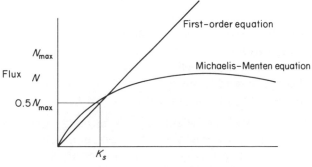

Fig. 12.3   Comparison of first-order and Michaelis–Menten models for describing organic flux in a fixed film reactor

The interesting difference to note between fixed-film and dispersed growth kinetics is that the former is concerned solely with organic transfer whereas the latter is concerned equally with substrate utilization and bacterial growth. Only in percolating filters does bacterial growth play an important role as noted in Section 12.4.

12.3 HIGH-RATE FILTERS

The situation in high-rate filters is relatively simple from a modelling viewpoint. High hydraulic loadings are used which permit the use of continuous dosing thus simplifying the description of the flow pattern. These high hydraulic loadings also control the growth of the bacterial film. It is therefore only necessary to consider organic transfer without any biological complications.

Organic transfer in high-rate filters is generally represented as a combination of diffusion limited transfer from the liquid layer and adsorption and reaction in the bacterial slime. In mathematical terms this is given by equation (12.11).

$$C_e = C_0 \exp\left(-\frac{K_m D}{Q}\right) \tag{12.11}$$

where

C_e = BOD of effluent
C_0 = BOD of influent
D = depth
Q = flow

and

$$\frac{1}{K_m} = \frac{1}{K_L A} + \frac{\alpha}{KX}$$

K_m = overall mass transfer coefficient
K_L = liquid film mass transfer coefficient
A = specific surface
α = specific adsorption coefficient
K = biological rate constant
X = dissolved oxygen concentration

The liquid mass transfer coefficient (K_L) is primarily dependent upon the Reynolds Number (surface irrigation rate), Schmidt Number (diffusivity of organic food) and the specific surface (A). The substrate removal is described by an adsorption coefficient (α) and the bacterial conversion is a first-order reaction with respect to the dissolved oxygen concentration.

High-rate filters have been found to be particulary sensitive to temperature changes. The form of the relationship is given by

$$K_T = K_{15} C^{(T-15)}$$

where

C = a constant
T = filter temperature $^\circ C$

12.4 PERCOLATING FILTERS

There is no really satisfactory model of percolating filters. The hydraulic and biological complexities are much greater than any of the other processes used in waste water treatment (except stabilization ponds).

Two fundamentally different approaches have been tried empirical and ecological. These are considered separately below:

12.4.1 Empirical filter models

Most of these models are based on the idea of the organic transfer as a first order reaction. The BOD removal is a function of retention time and assuming that the filter behaves like a plug-flow reactor the removal is therefore proportional to depth. This gives the following relationship

$$\frac{L_D}{L_0} = e^{-KD} \tag{12.12}$$

where

L_0 = Initial BOD,
L_D = BOD remaining at depth D,
K = biological rate coefficient
D = depth

This basic approach has been modified in various ways mainly

(a) to allow for a decrease in oxidisability with increasing retention time, and
(b) to increase the flexibility of the relationship for retention time;
(c) to allow for the decrease of active area with increasing depth.

Using these modifications the relationship becomes:

$$\frac{L_e}{L_0} = \frac{1}{1 + (CD^m/Q^n)} \tag{12.13}$$

where C, m and n are constants and Q = flow.

12.4.2 Ecological filter model

The physical conditions in a percolating filter are significantly different from those in a high-rate filter. The most notable differences are as follows:

1. The percentage voids of the support medium is much lower 40–45% compared with 90–99% of plastic packing. The flow pattern through the filter is therefore much more susceptible to blockage due to the growth of bacteria.
2. The lower hydraulic loadings employed in percolating filters necessitates the use of intermittent dosing. As a consequence, the flow pattern is more complex consisting of a series of quiescent intervals separated by short turbulent dosing periods.
3. Due to the lower hydraulic loading rates and the relatively high concentration near the top of a percolating filter, film growths are not controlled by surface erosion. Instead, the film growth in filters is regulated by a balance between bacterial growth rate and the rate of predation by the grazing fauna. This balance is generally successful in controlling film growth at hydraulically acceptable levels during the summer but lower winter temperatures differentially depress the grazing fauna and the winter period is characterized by build up of slime and consequently impaired flow (i.e. short circuiting).

These comments give some indication of the complexity of the percolating filter. In particular, the interrelationship between bacteria, grazing fauna, and the flow pattern is difficult to simulate. However, the understanding of the filter processes has reached a stage where a conceptual simulation model may be attempted.

The structure of the model is illustrated in Fig. 12.4. The flow of liquid through a filter is regarded as pulsating so that the overall retention time may be divided into a number of quiescent intervals separated by very brief vertical displacements. The number of such intervals is given by Retention time/Periodicity of dosing. In the majority of cases this will not be an integer so that the last quiescent period is shorter than the others. The transfer of organic matter is represented as taking place from liquid bacterial film during the quiescent periods. The transfer is by molecular diffusion so the process may be represented by Ficks' Law as:

$$\frac{\partial^2 c}{\partial x^2} = \frac{1}{D} \frac{\partial c}{\partial t} \tag{12.14}$$

with the initial condition

$$c(x, 0) = c_0 \qquad 0 < x < L$$

and the boundary conditions

$$c(0, t) = 0 \quad 0 < t < T$$

$$\frac{\partial c}{\partial t} (L, t) = 0 \quad 0 < t < T$$

where

c = concentration of organic matter
x = depth in the liquid layer with a maximum of L
D = diffusion coefficient
t = time interval up to a maximum of T which is the dosing interval

The mean concentration (c_m) at $t = T$ is given by equation (12.15)

$$c_m(T) = \frac{1}{L} \int_0^L c(x, T) \, \mathrm{d}x \tag{12.15}$$

which can be integrated to

$$c_m = c_0 \frac{8}{\pi^2} \sum_{n=0}^{\infty} \frac{e^{-DT(2n+1)^2\pi^2/4L^2}}{(2n+1)^2} \tag{12.16}$$

Hence the mean transfer per cycle (T) per unit area is:

$$(c_0 - c_m)L = c_0 L \left(1 - \frac{8}{\pi^2} \sum_{n=0}^{\infty} \frac{e^{-DT(2n+1)^2\pi^2/4L^2}}{(2n+1)^2} \right) \tag{12.17}$$

by substitution of

$$A = DT\pi^2/4L^2$$

a series solution can be obtained as shown in equation (12.18). This series truncates rapidly and sufficient accuracy can be obtained from two to three terms.

$$c_m = c_0 \frac{8}{\pi^2} e^{-A} + \frac{e^{-9A}}{9} + \frac{e^{-25A}}{25} + \ldots \tag{12.18}$$

During these periods, the rate of transfer decreases due to reduction in concentration and the formation of concentration gradients. In the displacement movements, mixing occurs so that the concentration is once again uniform. Therefore more frequent dosing tends to maximize the organic transfer.

The organic matter transferred to the bacterial film provides a food source for these organisms. The relationship between film growth, food concentration, and number of predators may be summarized as follows:

(a) Bacterial growth

$$S \frac{\mathrm{d}N}{\mathrm{d}t} = \frac{1}{N} \frac{K_c}{1+c} \tag{12.19}$$

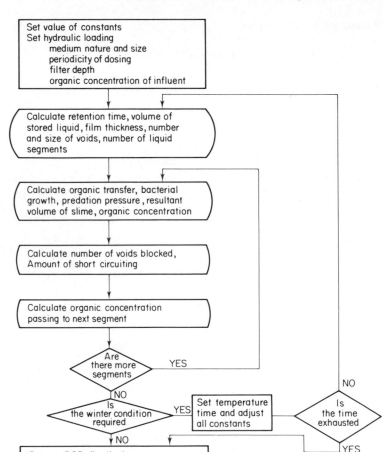

Fig. 12.4 Flow diagram of filter model

where

    N = Bacterial numbers,
    S = conversion factors for slime volume, (i.e. volume of slime associated with each bacterial cell)
    K = Michaelis–Menten constant
    c = concentration of organic matter

(b) Washout

$$\frac{dW}{dt} = SN^{K_2}(K_3 Q) \tag{12.20}$$

where

K_2 = washout constant which takes into account of decreased stability of large slime volume

K_3 = washout constant relating removal to hydraulics

Q = hydraulic loading

(c) Predation

$$\frac{\mathrm{d}P}{\mathrm{d}t} = K_4 SN \qquad (12.21)$$

where

P = numbers removed by predators

K_4 = conversion constant

The interaction of these processes is very dependent on temperature so that two conditions have to be defined

(i) Summer

$$\frac{\mathrm{d}P}{\mathrm{d}t} = \frac{\mathrm{d}N}{\mathrm{d}t} - \frac{\mathrm{d}W}{\mathrm{d}t} \qquad (12.22)$$

so that a balance slime volume exists.

(ii) Winter

$$\frac{\mathrm{d}N}{\mathrm{d}t} < \frac{\mathrm{d}P}{\mathrm{d}t} + \frac{\mathrm{d}W}{\mathrm{d}t} \qquad (12.23)$$

so that the slime volume steadily increases.

The effect of increased slime volume is felt to cause some shortcircuiting of liquid through that layer. The relationship between slime volume and short-circuiting is not well defined but is of the following form:

$$B = (S - \Delta S)K_5 \qquad (12.24)$$

where

ΔS = minimum volume at which bypassing occurs (this is a function of the type and size of filter medium)

K_5 = bypass constant

These calculations are carried out for each layer in the filter in succession as shown in Fig. 12.4 to give the BOD concentration at the top and bottom of each layer. For the summer condition a steady-state solution is obtained. For the winter condition, the equations are solved on a daily basis and this is iterated to give the solution at the required time.

The model described above is only at the development stage and cannot, as yet, be used for design or operation.

12.5 BIBLIOGRAPHY

Atkinson, B. (1974). *Biochemical Reactors*. Pim Ltd., London.
Kornegay, B. H. (1975). Modelling and simulation of fixed film biological reactors for carbonaceous waste treatment. Chapter 8. In: *Mathematical Modelling for Water Pollution Control Processes* (ed. McKeinath and Wanielista), Ann Arbor Science, Ann Arbor, Mich.
James, A. (1978). An ecological model of a percolating filter. Chapter 7. In: *Mathematical Models in Water Pollution Control*. Edited by A. James. J. Wiley & Sons, Chichester.
Ames, W. F., Behn, V. C. and Collings, W. Z. (1962). Transient operation of trickling filters, *Proceedings of American Society of Civil Engineers*. **88** (SA3), 21–38.

APPENDIX

Listing of the program for a model of a percolating filter

```
1 REM MODEL FOR PERCOLATING FILTER
4 DIM X(20), O(20)
10 INPUT 09
15 C0 = 09
20 READ H,E,R,S1,R1,F,D,Y
30 DATA 0.02.2.1200.40.1.0.2.0.004.03
40 REM H = HYDRAULIC LOADING (M3/M3 HR)
45 REM E = DEPTH (M)
50 REM R = CROSS SECTIONAL AREA(M2)
55 REM S1 - SPECIFIC SURFACE (M2/M3)
60 REM R1 = RETENTION TIME (HR)
63 REM F = DOSING INTERVAL (HR)
66 REM D = DIFFUSION COEFFICIENT (G/M2 HR)
68 REM Y = YIELD COEFFICIENT (G/G)
70 Z = R1/F
72 FOR I = 1 TO Z
73 READ X(I)
74 DATA 5,4,3,2,1
75 NEXT
80 S2 = S1*E/Z
90 V = R1*H/Z
100 T = V/S2*100
103 REM M = FILTER TEMPERATURE
106 M = 30
110 Y = Y*(1.05)↑(M − 20)
120 A = D*10*F/4*(T)↑2
130 REM K2 = RATE OF FILM REMOVAL
140 K2 = .01*(1.05)↑(M − 20)
```

```
150 FOR N = 1 TO 10
160 FOR B = 1 TO Z
170 C1CO*.85*(2.7)↑(−A)
180 PRINT "C1−", C1
200 X(B) = X(B) + X(B)*Y*(C0 − C1)*H − K2*X(B)
205 PRINT"X(B) − ",X(B)
210 V1 − 10*(5 − X(B))/100
215 PRINT V1
230 C(B) = C1*(1 − V1)*CO*V1
240 PRINT "EFF CONC = ",C(B),"B = ",B
250 C0 = C(B)
255 X1 = 1
260 NEXT B
262 STOP
265 C0 − C9
270 NEXT N
280 END
READY.
```

Chapter 13

The Modelling of Anaerobic Processes Used in Wastewater Treatment

A. JAMES

13.1 INTRODUCTION

There are a variety of anaerobic processes used in treating wastewaters, mainly digestion but also filters and contact digesters. In modelling these processes, the hydraulic situation needs to be considered first since all types of reactors from plug flow to completely mixed may be used. Once the hydraulics have been adequately described either from theoretical considerations or tracer studies, then attention can be concentrated on the microbiology and biochemistry. The approach to this is somewhat different to that used in aerobic processes. It is described in detail below. But before proceeding with this it is worthwhile examining in general terms the aims and requirements of anaerobic modelling.

Anaerobic processes in particular digestion, have been used for many years mainly for the treatment of sludges. Modelling has been used for simulating the digestion as a semi-quantitative guide to predict and avoid digester failure. With recent developments in the use of digestion, especially contact digestion, for treating industrial wastes the use of modelling has been extended and the need for more quantitative predictions has become more pressing. The modelling of anaerobic processes has not yet reached the stage where designs may be based directly on the models but useful guidelines can be obtained, and considerable assistance can be obtained with the operation of digesters.

13.2 THEORETICAL CONSIDERATIONS

The anaerobic breakdown of complex organic wastes may be divided into a number of stages as shown in Fig. 13.1.

The first stage is not always present and is chiefly associated with anaerobic breakdown of agricultural slurries or other complex waste materials. Where it does occur it is the rate-limiting step since it relies on outward movement of extra-cellular enzymes and the return movement of soluble organics. Where

Insoluble complex organics	Solub le organics (e.g. Monosaccarides)	Volatile acids	Methane and carbon
(e.g. Polysacchardies)	(Amino acids)		dioxides
(Proteins) (Lipids)	(Fatty acids)		

Fig. 13.1 Simple representation of anaerobic digestion as a three-stage process

the first step occurs it is obviously important to represent this in any simulation. But fortunately in many cases the organics present are largely in a soluble form and the first stage may be omitted from the model.

The modelling of the remaining two stages is based upon their relative rates. The decomposition of soluble organics is brought about by heterotrophic bacteria whch are numerous and active compared with the methane bacteria that bring about the decomposition of volatile acids. The latter may therefore be regarded as the rate limiting and the whole digestion process may be represented by this last stage. Most models of digestion therefore use the volatile acid equivalent of the soluble organic carbon as input. The conversion is regarded as so rapid compared to the methanogenesis that it is instantaneous.

Representation of methanogenesis is based on Monod kinetics with equations describing the growth of the methane bacteria and relating this to substrate removal. Some modification is needed to take account of inhibition at high concentrations of volatile acids as shown in Fig. 13.2.

The resulting growth function may be represented by:

$$\mu = \frac{\mu_{\max}}{1 + (K_s/S) + (S/K_i)} \tag{13.1}$$

in which μ = specific growth rate, time^{-1}; $\mu_{\max}$ = maximum specific growth rate in the absence of inhibition, time^{-1}; S = concentration of substrate, mass per unit volume; K_s = lower concentration of substrate giving a growth rate which is half the maximum, mass per unit volume; K_i = higher concentration. It can be seen that equation (13.1) is quadratic in form and will therefore give two

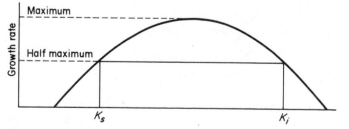

Fig. 13.2 Growth response of methanogenic bacteria to various concentrations of volatile acids

solutions corresponding to a particular growth rate. The higher range represents unstable conditions since any fluctuations in substrate concentration will not cause a homeostatic response (see Chapter 4 for explanation).

The degree of inhibition can vary. This can be represented by changing the value of K_i. As K_i tends to infinity then the inhibition tends to zero. It will be noted that even when K_s and K_i are well separated there is a significant reduction in the maximum specific growth rate attainable compared to the case without inhibition. The maximum rate attainable from $d\mu/dt = 0$ which gives

$$\text{Maximum attainable } \mu = \frac{\mu_{max}}{1 + 2[K_s/K_i]^{0.5}} \qquad (13.2)$$

with a corresponding substrate concentration S_m given by

$$S_m = (K_s K_i)^{0.5} \qquad (13.3)$$

The reasons for the inhibition are not clear but it appears to be connected with the concentration of unionized volatile acid. This is a function of both substrate and total acid concentration as shown in Fig. 13.3.

Since pH is a logarithmic function this principally determines the concentration of unionized acid. The inhibition function therefore needs to be modified to incorporate the effect on pH. This may be accomplished by considering the unionized acid as the limiting substrate and expressing K_s and K_i as concentrations of unionized substrate, thus

$$\mu = \frac{\mu_{max}}{1 + (K_s/HS) + (HS/K_i)} \qquad (13.4)$$

where HS—concentration of unionized substrate, mass per unit volume. The concentration of unionized substrate may be calculated from the mass action equilibrium

$$K_a = \frac{[H^+][S^-]}{[HS]} \qquad (13.5)$$

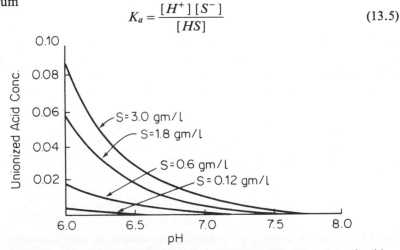

Fig. 13.3   Variation of unionized acid concentration with pH and total acid concentration (from Andrews, 1968)

where

$[S^-]$ = concentration of ionized substrate:
$[H^+]$ = concentration of hydrogen ions

and

K_a = ionization constant. This can be arranged to give

$$[HS] = \frac{[S^-][H^+]}{K_a} \tag{13.6}$$

For acetic acid the value of K_a is $10^{-4.5}$ at an ionic strength of 0.2 molar and a temperature of 38°C. If the pH is six or above, the total substrate concentration $[S]$ is approximately equal to the ionized substrate concentration $[S^-]$. There the maximum attainable growth rate may be calculated from

$$\mu = \frac{\mu_{max}}{1 + \frac{K_s K_a}{[S^-][H^+]} + \frac{[S^-][H^+]}{K_i K_a}} \tag{13.7}$$

and similarly the substrate concentration corresponding is given by

$$S_m = \frac{K_a}{[H^+]}(K_s K_i)^{0.5} \tag{13.8}$$

K_s and K_i are expressed as concentrations of unionized substrate and S_m is in units of total substrate concentration.

13.3 MODEL OF A DIGESTER

As pointed out in the introduction, the first step in modelling is to consider the hydraulic behaviour of the digester. Fortunately many digesters approximate closely to completely mixed reactors. Using this assumption the basic equations may be written as follows:

$$\frac{dX_2}{dt} = \frac{Q_1}{V} * X_1 - \frac{Q_2}{V} * X_2 + \frac{\mu_{max}}{1 + \frac{K_s}{HS_2} + \frac{HS_2}{K_i}} X_2 \tag{13.9}$$

$$\frac{dS_2}{dt} = \frac{Q_1}{V} * S_1 - \frac{Q_2}{V} * S_2 - \frac{1}{Y} \frac{\mu_{max}}{1 + \frac{K_s}{HS_2} + \frac{HS_2}{K_i}} X_2 \tag{13.10}$$

where

X_1 = concentration of methane bacteria in influent
X_2 = concentration of methane bacteria in effluent
S_1 = concentration of total volatile acids in influent
S_2 = concentration of total volatile acids in effluent
HS_2 = concentration of unionized volatile acids in effluent

HS_2 is calculated from the pH and S_2 using

$$HS_2 = \frac{[H^+][S_2^-]}{K_a}$$

For steady-state situations

$$\frac{dX_2}{dt} = \frac{dS_2}{dt} = 0$$

and assuming that there are no methane bacteria in the influent the steady-state values for bacteria and substrate are given by:

$$X_2 = Y(S_1 - S_2) \tag{13.11}$$

$$S_2 = \frac{\frac{K_iK_a}{[H^+]}\left(\mu_{max}\frac{Q}{V} - 1\right)\left\{\left[\frac{K_iK_a}{[H^+]}\right]^2\left(\mu_{max}\frac{Q}{V} - 1\right)^2 - 4K_sK_i\left[\frac{K_a}{[H^+]}\right]^2\right\}^{0.5}}{2} \tag{13.12}$$

As previously pointed out, two values for S_2 can be obtained from the above equation but only the lower value has any practical significance since it would not be feasible to operate at the higher range.

Figure 13.4 shows plots of the steady-state substrate concentration, S_2, for difference retention times at a constant pH. It can be seen that the retention time at which washout occurs increases as the substrate becomes more inhibitory. The retention time at which washout occurs may be calculated as

$$\left(\frac{V}{Q}\right)_w = \frac{1}{\mu_{max}}\left[1 + 2\left(\frac{K_s}{K_i}\right)\right]^{0.5} \tag{13.13}$$

The effect of pH on the steady-state substrate concentration is shown in Fig. 13.5. At a constant value for K_i the retention time at which washout occurs

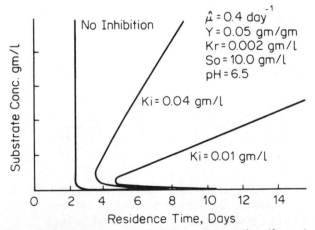

Fig. 13.4   Effect of retention time on substrate concentration (from Andrews, 1968)

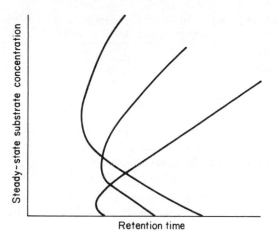

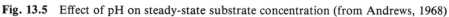

Fig. 13.5   Effect of pH on steady-state substrate concentration (from Andrews, 1968)

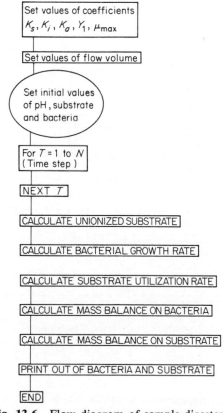

Fig. 13.6   Flow diagram of sample digester model

Table 13.1   Range of reported values of kinetic coefficients for methomogenic bacteria at 35°C

Substrate	μ_{max}	Y	K_s	K_d
Acetate	0.324	0.04	154	0.019
Proprionate	0.318	0.027	39	0.01
Butyrate	0.389	0.017	7	0.027

is not affected by changing pH. But the pH causes the equilibrium concentration of substrate to increase with increase in pH because the unionized acid concentration is lower with increase in pH values.

The flow diagram for the model is presented in Fig. 13.6. Initial conditions need to be specified in terms of bacterial concentration, substrate concentration and pH. The choice of these can be critical to the output of the model

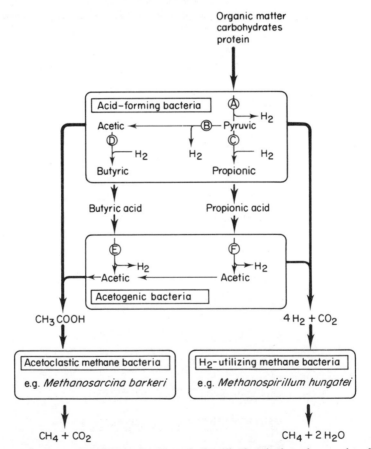

Fig. 13.7   Diagrammatic representation of the biochemical pathways involved in anaerobic digestion (from Mosey, 1982)

because unlike activated sludge models the whole range of possible combinations does not cause homeostasis. It is therefore best to calculate the steady-state concentrations from the above equations and to use these, or some close approximations as the initial values. A listing of the model is given in the Appendix to this chapter.

Values are also required for the model coefficients. In the absence of particular data from laboratory studies, the data in Table 13.1 may be useful.

The structure of the model is a matter of choice, depending to a large extent on the degree of complexity required. Figure 13.6 shows a simple digester model in which only pH, volatile acids and methane bacteria are considered. This may be elaborated to take account of other important processes e.g.

(a) Other factors affecting the pH-alkalinity relationship such as carbonate–bicarbonate equilibrium and ammonia concentration.

(b) The influence of the oxidation-reduction potential on methanogenesis.

(c) The effect of the gas phase—CO_2, H_2 and CH_4 composition and rate of gas production on the process.

Recent work on the microbiology of anaerobic breakdown has indicated that a number of different groups and pathways are involved as shown in Fig. 13.7. More complex models will therefore be required to fully describe the process but these will also require more data on growth rates, yields, etc.

13.4 BIBLIOGRAPHY

Andrews, J. F. (1968). A dynamic model of the anaerobic digestion process, *Proceedings of 23rd Industrial Wastes Conference, Purdue*, p. 285. University of Purdue.

Lawrence, A. W. and McCarthy, P. L. (1969). Kinetics of methane fermentation in anaerobic treatment, *Journal of Water Pollution Control Federation*. R1–R17.

Mosey, F. E. (1982). New Developments in the Anaerobic Treatment of Industrial Wastes, *Journal of the Institute of Water Pollution Control* **81**, 540–552.

Hungate, R. E. (1970). Formate as an Intermediate in the Bovine Rumer Fermentation. *Journal of Bacteriology*. **102**, 1–2, 389.

Andrews, J. F. (1978). The Development of a Dynamic Model and Control Strategies for the Anaerobic Digestion Process. Chapter 13. In: *Mathematical Models in Water Pollution Control* (ed. A. James). John Wiley & Sons, Chichester.

APPENDIX

Listing of a program for a digester model (simple version)

```
10 REMMODEL OF ANAEROBIC DIGESTER
20 REM   D = DILUTION RATE (DAY − 1)
30 REM   Y = YIELD OF METHANE BACTERIA
40 REM XO = INPUT CONC OF METHANE BACTERIA (G/M3)
```

```
 50 REM  X1 =  DIGESTER CONC OF METHAN BACTERIA (G/M3)
 60 REM  SO =  INPUT CONC OF SUBSTRATE (G/M3)
 70 REM  S1 =  DIGESTER CONC OF SUBSTRATE (G/M3)
 80 REM  K1 =  KS
 90 REM  K2 =  KI
100 REM  K3 =  KA
105 REM   H =  HYDROGEN ION CONCENTRATION
110         D  =  .1
120         Y  =  .05
130        SO  =  10000
140        XO  =  0
150        X1  =  50
160        S1  =  100
170        K3  =  .000001
180         H  =  .000001
190        K2  =  40
200        K1  =  2
210         U  =  .4
220        FOR T  =  1  TO  10
230        S2  =  S1*H/K3
240        SE  =  S2 + 2
250        S4  =  S3/K2
260        F  =  S2/(S4 + S2 + K1)
270        U1  =  U*F
280        X2  =  D*XO* – D*X1 + U1*X1
290        S5  =  (U1*X1)/Y
300        S6  =  D*SO – D*S1 – S5
310        X1  =  X1 + X2
320        S1  –  S1 + S6
330        PRINT "S1 = ", S1, "X1  =  ", X1
340        NEXT T
350        END
```

APPENDIX

Listing of digester model (improved version)

```
10 REM DIGESTER MODEL
20 REM V = DIGESTER VOLUME(M3)
30 V = 10000
40 REM U = MAXIMUM GROWTH RATE(D – 1)
50 U = 0.4
60 REM Y – YIELD(DIMENSIONLESS)
```

```
 70  Y = 0.05
 80  REM K1 = MICHAELIS MENTEN COEF(G/M3)
 90  K1 = 20
100  REM K2 = INHIBITION COEF(G/M3)
110  K2 = 200
120  REM K3 = IONISATION COEF(G/M3)
130  K3 = 0.000001
140  REM PH = HYDROGEN ION CONC(G/M3)
150  PH = 0.000001
160  REM S1 = INFLUENT SUBSTRATE CONC(G/M3)
170  PRINT"INPUT S1"
180  INPUT S1
190  REM S2 = SUBSTRATE CONC IN DIGESTER
200  S2 = 100
210  REM X2 = BACTERIAL CONC IN DIGESTER
220  X2 = 50
230  REMQ = HYDRAULIC LOADING(M3/D)
240  PRINT"INPUT Q"
250  INPUT Q
260  REM DAILY LOOP
270  FOR H = 1 TO 24
280  REM HOURLY LOOP
290  FOR H = 1 TO 24
300  REM CALC OF GROWTH RATE
310  S3 = S2*(PH/K3)
320  S4 = S3↑2
330  S5 = S4/K2
340  F = S3/(S5 + S3 + k1)
350  U1 = U*F
360  REM CALC OF BACTERIAL BALANCE
370  X3 = (U1*X2) − (Q*X2/V)
380  REM CALC OF SUBSTRATE BALANCE
390  S6 = (Q*S1/V) − (Q*S2/V) − (U1*X2/Y)
400  REM CALC OF NEW BACTERIAL CONC
410  X2 = X2 + (X3/24)
420  REM CALC OF NEW SUBSTRATE CONC
430  S2 = S2 + (S6/24)
440  REM END OF HOURLY LOOP
450  NEXT H
460  PRINT"SUBSTRATE = ", S2
470  PRINT"BACTERIA = ",X2
480  REM END OF DAILY LOOP
490  NEXT D
500  END
```

Notes on digester models

The main assumption are as follows:

(a) The rate limiting step is methanogenesis.
(b) The rate of growth of the methanogenic bacteria is limited by the concentration of unionized volatile acid.
(c) The pH of the digester is being controlled.

Both models use the same kinetics but the second version has been improved by the addition of more explanatory comments.

Chapter 14

The Modelling of Overall Treatment

A. James

14.1 INTRODUCTION

There are a number of attractions to models of overall treatment. On the design side they overcome the problem of interfacing units, so that changes in the performance of a primary unit are passed on to secondary and tertiary units enabling a wide range of possibilities to be explored.

This enables optimization to be performed on the combined units so that the cheapest overall scheme can be chosen. On the operational side they enable the simulation of shock loads or special operating conditions (e.g. primary clarifier out of action) to be carried out to determine the best operating procedure or the likely period of recovery.

There are, however, considerable difficulties in formulating such models. The understanding of process relationship for individual stages is often inadequate especially outside the normal range of operation. An example of this is shown in Fig. 14.1 pointing to the tenuous nature of the relationship at short retention times.

In such circumstances it seems desirable to use an interactive approach so that the user is able to explore a large number of permutations but at the same time retains control over the possibilities being explored. This avoids the

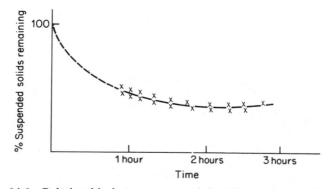

Fig. 14.1 Relationship between suspended solids and retention time

danger of producing solutions from a sophisticated optimization model that are unacceptable from an engineering standpoint.

One further area of uncertainty is that relating to costs. These need to be introduced if alternative solutions are to be compared on an economic basis. But it is difficult to translate costs across space and time. A better approach is therefore to include the sub-routine for calculating costs and leave it to the designer to input the cost data that he feels are appropriate for his own conditions. The following notes describe both approaches—an optimization model and an interactive model.

14.2 OPTIMIZATION MODEL

14.2.1 Inputs to the model

Although it would be desirable to formulate a model which is capable of handling many different quality parameters such as NH_3, PO_4, coliforms, etc., it is unrealistic at the moment to handle more than BOD and suspended solids. Even the relationship between these two parameters changes, for example, in primary sedimentation, BOD removal is $30 - 40\%$ compared with $60 - 70\%$ removal of suspended solids. It is, therefore, necessary to choose a parameter on which to base the optimisation and then check the optimum design with the other parameter(s).

Cost data can be entered into the model in various ways. The simplest form is graphical as shown in Fig. 14.2 relating the cost to the volume by relating the performance to the retention time. It is, however, much more accurate to break down the costs as far as possible to individual operations like excavation, shuttering, etc. and to compute the cost of units by summing the costs of all the operations. The first method is suitable for calculating comparative costs but the latter method (if data are available) can be used to give absolute costs on a world-wide basis.

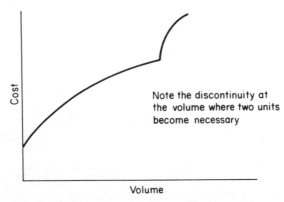

Fig. 14.2   Cost/volume relationship

14.2.2 Simulation model

The choice of the process simulations to use is up to the individual designer. In some models of overall treatment the individual stages are represented by the explicit, deterministic models described in Chapters 10–13. But other models use a black-box approach with empirical relationships derived on a purely statistical basis. The following example uses black-box models to represent a simple treatment combination of primary sedimentation and activated sludge. This is illustrated in Fig. 14.3a. The performance of the two sedimentation tanks may be represented by the following equations:

$$P = (0.0004S)(1 - e^{-0.7T}) \tag{14.1}$$

$$B = (0.9 - 0.5P)B_1 \tag{14.2}$$

where

P = proportion of solids removed
S = concentration of solids removed
B = concentration of BOD remaining
T = retention time = V/Q
V = volume of sedimentation tank
Q = flow
B_1 = influent BOD

These equations describe the performance of a primary clarifier reasonably well but are less suitable for describing secondary settlement since the removal of solids and BOD is so dependent upon sludge age. The performance of the aeration tank may be simply represented by

$$B = 10 + (L - 10) \exp\left(-\frac{CTm}{2000}\right) \tag{14.3}$$

where

L = initial concentration of BOD
B = concentration of BOD remaining
C = temperature coefficient

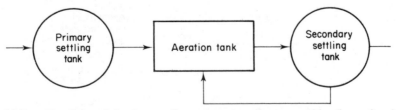

Fig. 14.3a Simple model of overall treatment using a combination of primary sedimentation and activated sludge

$$= 0.0073t^2 - 0.0827t + 0.7$$

t = temperature °C

T = retention time

m = concentration of mixed liquor suspended solids

This equation is designed to give a minimum residual BOD of 10 regardless of conditions in the aeration tank. It also assumes that the aeration unit is operating at steady-state i.e. at a fixed MLSS.

14.2.3 Optimization model

The optimization model is formulated to find the lowest cost solution using the above combination of units to achieve a fixed effluent quality. The model is represented diagrammatically in Fig. 14.3b using BOD as the parameter. Once the flow of waste and the influent and effluent BODs are specified as the boundary conditions then there are only two variables—the first and second intermediate BODs. The first intermediate BOD is determined by the retention time in the primary clarifier (i.e. by the volume). The second intermediate BOD is determined by the retention time in the aeration tank (for a fixed level of MLSS) and this also determines the size of the final clarifier.

The optimization procedure is to choose pairs of values for first and second intermediate BOD's and using equations (14.1) to (14.3) to determine the sizes of the three units. These sizes can then be translated into costs by relationships like the one shown in Fig. 14.2. So the total cost, C_T can be calculated from:

$$C_T = f_1(V_1) + f_2(V_2) + f_3(V_3)$$

where

V_1, V_2 and V_3 are the costs of the individual units

and

f_1, f_2 and f_3 are the functions relating the costs to the volume

It can be seen from Fig. 14.2 that the cost/volume relationship is unlikely to

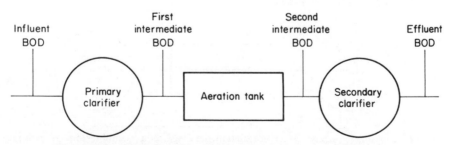

Fig. 14.3b Diagrammatic representation of optimization model

be linear. The optimization technique employed must therefore be capable of handling non-linear relationships. The costs functions give a unimodal cost surface allowing the use of a gradient method.

One method of solving this problem is to use non-linear simplex. The flow diagram for the solution is shown in Fig. 14.4 and the program is given in the Appendix.

The program may be run more than once as shown in Fig. 14.5 which shows the runs superimposed upon the cost surface. The simplex strategy is to choose the worst (i.e. most expensive) of the three designs and perform a reflexion.

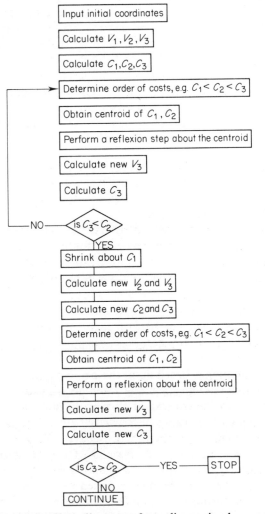

Fig. 14.4 Flow diagram of non-linear simplex model

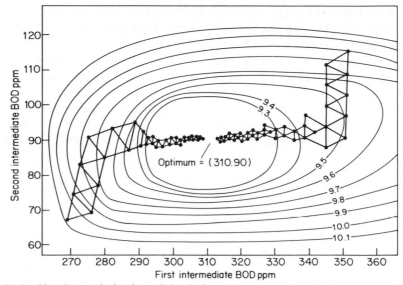

Fig. 14.5   Simplex optimization of the design of a waste-water treatment plant (from Knapton, 1978)

This process continues until the reflexion yields a design which is still the worst whereupon the designs are shrunk about the best (i.e. cheapest) design. The process then continues with the shrunken triangle until reflexion or further shrinking cannot give a better design.

The same problem may be tackled by dynamic programming. Using this method, the problem is first simplified to a finite number of alternatives by constraining the first and second intermediate BOD's to only a few possibilities, e.g. three values each. This is represented diagrammatically in Fig. 14.6 which shows the problem as a network. The influent enters at A and the effluent leaves at H and any of the nine possible paths connecting these may be the cheapest solution. The numbers on each path represent the cost of using that path. The total cost of any design is thus given by the sum of its constituent paths.

Dynamic programming is based upon the principle that, 'If a series of decisions forms an optimal design, then any given decision affects only those decisions which follow. In the context of this problem, it means evaluating the solutions in the network working backwards from H. Once E, F or G have been reached the remainder of the design is fixed. Once B is reached there is a choice of subsequent routes via E, F or G of which the route via E is the cheapest. The same argument applies to C and D such that if C is reached the optimal route is $C \rightarrow F \rightarrow H$ and if D is reached the route is $D \rightarrow E \rightarrow H$. From A it is possible to go through B, C or D. Whichever link is made, the rest of the route has previously been determined ($B \rightarrow E \rightarrow H$ or $C \rightarrow F \rightarrow H$ or

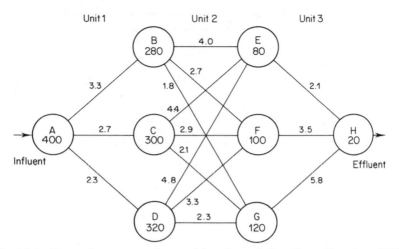

Fig. 14.6 Dynamic programming problem formulation (from Knapton, 1978)

$D \to E \to H$). Thus, if the costs of the links from A are added to the running totals at B, C or D, it becomes apparent that the optimal route is $A \to D \to E \to H$.

Dynamic programming operates by considering the design process as a sequential operation and allows the optimal design to be calculated without every feasible design being explored. Its chief advantage is that it does not require a unimodal cost surface but it can operate only on the sequential type of problem.

14.3 INTERACTIVE DESIGN MODEL

The philosophy of this approach is fundamentally different from that of the optimization model although the process design and cost calculation parts may be identical. The interactive model aims at involving the designer in the decision making process at all stages instead of relying on minimum cost algorithms.

The other main difference is that the interactive model provides a complete design including calculation of quantities, and costs and also preparation of drawings.

A flow diagram for an interactive model is presented in Fig. 14.7. It shows how the design is produced as the result of calculations by the computer and decisions by the designer. To avoid overcomplicating, the diagram Fig. 14.7 shows the process as occcuring on a 'once through' basis but in practice the designer may explore any number of combination of units, various sizes of units and various site layouts. These would normally be explored only as far as the calculation of cost and drawings would only be prepared for the best design.

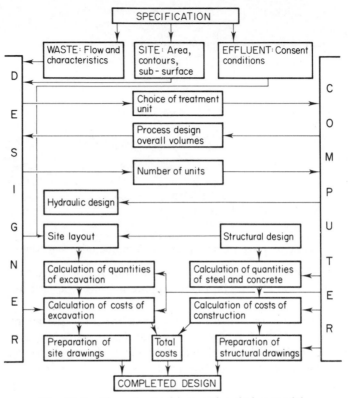

Fig. 14.7   Flow chart of interactive design model

14.4 BIBLIOGRAPHY

Ciria (1973). Cost effective sewage treatment, *The Creation of an Optimization Model,*
Report No. 46 of the Construction Industry Research and Information Assn.,
London.
Knapton, J. (1978). Optimization and its Application to a Unit Process Design Pro-
blem. Chapter 4. In: *Mathematical Models in Water Pollution Control* (ed. by A.
James). J. Wiley & Sons, Chichester.
Silverston, P.L. (1972). Simulation of the mean performance of municipal waste treat-
ment plants, *Water Research.* **6** (9), 1101 – 1112.

P.1 Non-linear simplex programme for optimization of sewage treatment

```
10 DIM   A(21), P(8), V(9),  C(9)
20 INPUT Q, S, B, M, E
30 REM INPUT INITIAL COORDINATES
40 FOR N = 1 to 6
```

```
 50 READ P(N)
 60 NEXT N
 70 REM EVALUATE EACH DESIGN
 80 FOR N = 1 TO 3
 90 B(N) = P(N)
100 V(W) = 0.7*Q*LOG(1 − B1 × (1.8 − 2*B(N))/.0004*S)
110 NEXT N
120 FOR N = 4 TO 6
130 L(N) = B(N)
140 V(N) = ((2000/(C*M)) * log((B2 − 10)/(L(N) − 10))) * Q
150 NEXT N
160 FOR N = 7 TO 9
170 V(N) = 0.7*Q*LOG (1 − B(N) * (1.8 − 2 × E/ .0004*S))
180 NEXT N
190 REM CALCULATE COSTS
200 C(1) = F1 * V(1) + F2 * V(4) + F3*V(7)
210 C(2) = F1 * V(2) + F2 * V(5) + F3*V(8)
220 C(3) = F1 * V(3) + F2 * V(6) + F3*V(9)
230 REM DETERMINE BEST AND WORST DESIGN
240 IW = 1
250 IB = 1
260 IF(VALUE (3).GT.VALUE (2))GO TO 290
270 IF(VALUE (2).GT.VALUE (1) IW = 2
280 GO TO 300
290 IF(VALUE (3).GT.VALUE (1)) IW = 3
300 IF(VALUE (3),LT.VALUE (2)) GO TO 330
310 IF(VALUE (2).LT.VALUE (1) =    IB
320 GO TO 340
330 IF(VALUE (3).LT.VALUE (1)) IB = 3
340 P = 0 : Q = 0
350 REM OBTAIN THE CENTROID OF ALL DESIGNS EXCEPT THE WORST
360 FOR I = 1,3
370 IF(I.PO.IW) GO TO 400
380 P = P + 0.5*P(I*2 − 1)
390 Q = Q + 0.5*P(I*2)
400 CONTINUE
410 REM PERFORM A REFLEXION STEP
420 P(7) = 2*P − P(IW*2 − 1)
430 P(8) = 2*Q − P(IW*2)
440 A(5) = P(7)
450 A(6) = P(8)
460 GO TO 70
470 VALUE (4) = A(21)
```

```
480 REM JUMP TO LABEL 11 IF THE REFLEXION WAS A SUCCESS
490 IF(VALUE (4).LT.VALUE(IWER) go to 600
500 REM SHRINK THE SIMPLEX ABOUT THE CURRENT BEST DESIGN
510 DO 10 I = 1,3
520 IF(I.BQ.LB) GO TO 550
530 P(1*2 - 1) = 0.5*(P(I*2 - 1) + P(IB*2))
540 P(I*2) = 0.5*(P(I*2) + P(IB*2))
550 CONTINUE
560 REM STEP COMPLETE
570 GO TO 10
580 REM REPLACE THE CURRENT WORST DESIGN BY THE REFLEXION
600 P(IW*2 - 1) = P(7)
610 P(IW*2) = P(8)
620 GO TO 10
630 REM STEP COMPLETE
640 END
```

Index